SYMBOLIC ANALYSIS IN ANALOG INTEGRATED CIRCUIT DESIGN

THE KLUWER INTERNATIONAL SERIES IN ENGINEERING AND COMPUTER SCIENCE

ANALOG CIRCUITS AND SIGNAL PROCESSING

Consulting Editor

Mohammed Ismail

Ohio State University

Related Titles:

SWITCHED-CURRENT DESIGN AND IMPLEMENTATION OF OVERSAMPLING A/D CONVERTERS, *Nianxiong Tan,* ISBN: 0-7923-9963-3

CMOS WIRELESS TRANSCEIVER DESIGN, *Jan Crols, Michiel Steyaert,* ISBN: 0-7923-9960-9

DESIGN OF LOW-VOLTAGE, LOW-POWER OPERATIONAL AMPLIFIER CELLS, *Ron Hogervorst, Johan H. Huijsing,* ISBN: 0-7923-9781-9

VLSI-COMPATIBLE IMPLEMENTATIONS FOR ARTIFICIAL NEURAL NETWORKS, *Sied Mehdi Fakhraie, Kenneth Carless Smith,* ISBN: 0-7923-9825-4

CHARACTERIZATION METHODS FOR SUBMICRON MOSFETs, edited by *Hisham Haddara,* ISBN: 0-7923-9695-2

LOW-VOLTAGE LOW-POWER ANALOG INTEGRATED CIRCUITS, edited by *Wouter Serdijn,* ISBN: 0-7923-9608-1

INTEGRATED VIDEO-FREQUENCY CONTINUOUS-TIME FILTERS: *High-Performance Realizations in BiCMOS, Scott D. Willingham, Ken Martin,* ISBN: 0-7923-9595-6

FEED-FORWARD NEURAL NETWORKS: *Vector Decomposition Analysis, Modelling and Analog Implementation, Anne-Johan Annema,* ISBN: 0-7923-9567-0

FREQUENCY COMPENSATION TECHNIQUES LOW-POWER OPERATIONAL AMPLIFIERS, *Ruud Easchauzier, Johan Huijsing,* ISBN: 0-7923-9565-4

ANALOG SIGNAL GENERATION FOR BIST OF MIXED-SIGNAL INTEGRATED CIRCUITS, *Gordon W. Roberts, Albert K. Lu,* ISBN: 0-7923-9564-6

INTEGRATED FIBER-OPTIC RECEIVERS, *Aaron Buchwald, Kenneth W. Martin,* ISBN: 0-7923-9549-2

MODELING WITH AN ANALOG HARDWARE DESCRIPTION LANGUAGE, *H. Alan Mantooth, Mike Fiegenbaum,* ISBN: 0-7923-9516-6

LOW-VOLTAGE CMOS OPERATIONAL AMPLIFIERS: *Theory, Design and Implementation, Satoshi Sakurai, Mohammed Ismail,* ISBN: 0-7923-9507-7

ANALYSIS AND SYNTHESIS OF MOS TRANSLINEAR CIRCUITS, *Remco J. Wiegerink,* ISBN: 0-7923-9390-2

COMPUTER-AIDED DESIGN OF ANALOG CIRCUITS AND SYSTEMS, *L. Richard Carley, Ronald S. Gyurcsik,* ISBN: 0-7923-9351-1

HIGH-PERFORMANCE CMOS CONTINUOUS-TIME FILTERS, *José Silva-Martínez, Michiel Steyaert, Willy Sansen,* ISBN: 0-7923-9339-2

SYMBOLIC ANALYSIS OF ANALOG CIRCUITS: *Techniques and Applications, Lawrence P. Huelsman, Georges G. E. Gielen,* ISBN: 0-7923-9324-4

DESIGN OF LOW-VOLTAGE BIPOLAR OPERATIONAL AMPLIFIERS, *M. Jeroen Fonderie, Johan H. Huijsing,* ISBN: 0-7923-9317-1

STATISTICAL MODELING FOR COMPUTER-AIDED DESIGN OF MOS VLSI CIRCUITS, *Christopher Michael, Mohammed Ismail,* ISBN: 0-7923-9299-X

SELECTIVE LINEAR-PHASE SWITCHED-CAPACITOR AND DIGITAL FILTERS, *Hussein Baher,* ISBN: 0-7923-9298-1

ANALOG CMOS FILTERS FOR VERY HIGH FREQUENCIES, *Bram Nauta,* ISBN: 0-7923-9272-8

ANALOG VLSI NEURAL NETWORKS, *Yoshiyasu Takefuji,* ISBN: 0-7923-9273-6

ANALOG VLSI IMPLEMENTATION OF NEURAL NETWORKS, *Carver A. Mead, Mohammed Ismail,* ISBN: 0-7923-9049-7

AN INTRODUCTION TO ANALOG VLSI DESIGN AUTOMATION, *Mohammed Ismail, José Franca,* ISBN: 0-7923-9071-7

SYMBOLIC ANALYSIS IN ANALOG INTEGRATED CIRCUIT DESIGN

by

Henrik Floberg
Lund University
Sweden

KLUWER ACADEMIC PUBLISHERS
Boston / Dordrecht / London

Distributors for North America:
Kluwer Academic Publishers
101 Philip Drive
Assinippi Park
Norwell, Massachusetts 02061 USA

Distributors for all other countries:
Kluwer Academic Publishers Group
Distribution Centre
Post Office Box 322
3300 AH Dordrecht, THE NETHERLANDS

Library of Congress Cataloging-in-Publication Data

A C.I.P. Catalogue record for this book is available
from the Library of Congress.

Copyright © 1997 by Kluwer Academic Publishers

All rights reserved. No part of this publication may be reproduced, stored in a retrieval system or transmitted in any form or by any means, mechanical, photo-copying, recording, or otherwise, without the prior written permission of the publisher, Kluwer Academic Publishers, 101 Philip Drive, Assinippi Park, Norwell, Massachusetts 02061

Printed on acid-free paper.

Printed in the United States of America

Contents

1 Introduction — 1
   1.1 Symbolic Analysis 1
   1.2 Motivation 6
   1.3 Book Outline 13

2 Network Theory — 15
   2.1 Transmission Function 15
   2.2 Poles and Zeros 16
   2.3 Nodes and Ports 17
   2.4 Network Elements 18
      2.4.1 Passive Elements 18
      2.4.2 Controlled Sources 19
      2.4.3 Nonlinear Elements 19
      2.4.4 Nullor 19
      2.4.5 Gyrator 20

3 Symbolic Analysis — 23
   3.1 Circuit Equations 23
      3.1.1 Tableau Formulation 25
      3.1.2 Mesh Formulation 26
      3.1.3 Loop Formulation 27

		3.1.4	Nodal Formulation. 28

- 3.1.4 Nodal Formulation. 28
- 3.1.5 Modified Nodal Formulation 29
- 3.1.6 Summary. 29
- 3.2 Symbolic Methods 30
  - 3.2.1 Signal-Flow Graph Method 30
  - 3.2.2 Tree Enumeration Method. 31
  - 3.2.3 Interpolation Method 32
  - 3.2.4 Parameter Extraction Method 33
  - 3.2.5 Determinant Methods 33
  - 3.2.6 Summary. 33
- 3.3 Admittance Matrix 34
- 3.4 Network Functions 35
- 3.5 Approximation Methods 37

4 Extended Pole-Splitting 39
- 4.1 Numerical Roots 39
- 4.2 Approximate Symbolic Roots 41
- 4.3 Two-Stage Operational Amplifier. 43
- 4.4 Single-Stage Operational Amplifier 45

5 Compacted Nodal Analysis 49
- 5.1 Transadmittances 49
- 5.2 Nullor. 51
- 5.3 Network Functions Using CNA 52
- 5.4 Block Diagram Elements 53

6 Nullor Synthesis 55
- 6.1 Two-Port Network 56
- 6.2 Basic Negative-Feedback Configurations 57

	6.3	Nullor Element	59
	6.4	Nonideal Transistor	61
	6.5	Transistor Implementations of the Nullor	62
	6.6	Transistor Amplifiers	65
	6.7	Alternative Transistor Amplifiers	68
7	**Transistor Models**		**75**
	7.1	Network Elements	76
	7.2	Nodal Formulation	76
	7.3	Nullor	77
	7.4	Small-Signal Transistor Models	78
	7.5	Approximate Symbolic Expressions	80
8	**Symbolic Distortion Analysis**		**83**
	8.1	Nonlinear Active Networks	84
		8.1.1 Basic Feedback Model	85
		8.1.2 Large-Signal Analysis	86
		8.1.3 Harmonic Balance	88
	8.2	Describing Functions	90
		8.2.1 General Feedback Model	91
		8.2.2 Nonlinear Function	92
		8.2.3 Fourier Transformation	92
		8.2.4 Harmonic Equations	94
		8.2.5 Harmonic Approximation	95
		8.2.6 Summary	96
	8.3	Examples	96
		8.3.1 Parabolic Function	97
		8.3.2 Exponential Function	101
		8.3.3 Saturation Function	105

9	**Switched-Capacitor Networks**	**109**
	9.1 Switched Capacitor	110
	9.2 Equivalent Analog Circuit.	111
	9.3 Switches and Operational Amplifiers	113
	9.4 Switched-Capacitor Amplifier.	114
	9.5 Switched-Capacitor Biquad	116
10	**CASCA**	**119**
	10.1 Language	120
	10.1.1 Symbolic Variables	120
	10.1.2 Symbolic Functions	121
	10.1.3 Miscellaneous Functions	122
	10.1.4 Polynomial Functions	122
	10.2 Circuit Description	122
	10.2.1 Circuit Elements	123
	10.2.2 Transadmittances	123
	10.2.3 Nullors	125
	10.2.4 Subcircuits	126
	10.2.5 Block Diagram Elements	127
	10.3 Graphics	127
11	**Conclusions**	**131**
A	**Determinants**	**135**
	A.1 Definition	135
	A.2 Minors and Cofactors	136
	A.3 Cramer's Rule	136
B	**Cubic Polynomial Equation**	**139**

C	**Describing Function Method**	**143**
	C.1 Variable Notation	143
	C.2 Numerical Values	144
D	**CASCA Examples**	**145**
	D.1 Active RC-Filter.	146
	D.2 Two-Stage Operational Amplifier	147
	D.3 SC-Biquad	152

Bibliography   **155**

Index   **163**

Preface

Computer aided symbolic circuit analysis is useful in analog integrated circuit design. Analytic expressions for the network transfer functions contain information that is not provided by a numerical simulation result. However, these expressions are generally extremely long and difficult to interpret and it is, therefore, necessary to be able to approximate them guided by the magnitude of the individual circuit parameters. Engineering has been described as "the art of making approximations". The inclusion of symbolic analysis in analog circuit design reduces the implied risk of ambiguity during the approximation process.

A systematic method based on the nullor concept is used to obtain the basic feedback transistor amplifier configurations. The extended pole-splitting technique makes it easier to obtain approximate expressions for the locations of poles and zeros for feedback amplifiers and gives insight into performance and stability. The choice of adequate transistor models for analog circuits is essential to the circuit designer. Fairly simple transistor models are sufficient to produce approximate expressions that are much more accurate than the numerical values available for physical active devices. Practical circuit design minimizes the influence of certain parasitics and validates the use of less refined models. An unusual feature in this book is the consistent use of the transadmittance with finite (linear or nonlinear) or infinite (i.e. nullor) gain as the only requisite circuit element. The describing function method is used to obtain approximate symbolic expressions for the harmonic distortion generated by a soft or hard transconductance nonlinearity embedded in an arbitrary linear network.

A tool (i.e. CASCA) for symbolic analysis in analog circuit design has been designed and implemented. The integrated graphics package makes it possible to compare the results produced by CASCA with the output from numerical circuit simulator programs. CASCA uses nodal analysis which has been extended to allow nullors (i.e. compacted nodal analysis) and block diagrams. The program is intended for analysis of time-continuous networks but may also be used to analyse time-discrete switched-capacitor networks.

Chapter 2 gives the reader basic knowledge about network theory and introduces some of the concepts found in the following chapters. Different formulation methods for linear circuit equations and some symbolic methods to solve these equations are presented in chapter 3. Chapter 4 presents the extended pole-splitting method which is used to obtain approximate symbolic expressions for the roots of a polynomial function. The compacted nodal analysis formulation is described in chapter 5. That the ideal nullor element can be used for transistor amplifier synthesis is shown in chapter 6. Chapter 7 presents different linear transistor models suitable for symbolic circuit analysis. In chapter 8, the describing function method is used for symbolic distortion analysis. The symbolic analysis of switched-capacitor networks using any tool intended for analysis of time-continuous networks is explained in chapter 9. Chapter 10 discusses a tool for symbolic analysis (i.e. CASCA) that has been implemented to study the ideas in this book. Chapter 11, finally, concludes the book.

Appendix A gives some basic determinant theory. The exact symbolic roots obtained with Maple for a cubic polynomial equation are given in appendix B. Appendix C contains the variable notation used for the describing function method in chapter 8. Some examples of circuit programs for CASCA are found in appendix D.

I would like to thank my advisor Sven Mattisson for his support during my work on *Symbolic Analysis in Analog Integrated Circuit Design* at the Circuit Design Laboratory, Department of Applied Electronics, Lund University, Sweden.

I would also like to thank my parents, friends, fellow graduate students and the staff at the Department of Applied Electronics for all their encouragement and help.

This work was supported by Apple Computer in Sweden and the Swedish National Board for Industrial and Technical Development (NUTEK).

Henrik Floberg

There is nothing so practical as a really good theory.
LUDWIG BOLTZMANN[1]

---

1. After Peter J. Baxandall, Wireless World, Jan. 1978.

Chapter 1

Introduction

1.1 Symbolic Analysis

The trend today is to move towards the integration of entire systems on a single chip. These application-specific integrated circuits (ASIC) include both analog and digital subsystems. Sophisticated tools such as silicon compilers are available for the design of the digital part but similar analog design tools are still fairly scarce. The analog part is often designed manually by experienced designers using hand calculations, iterative numerical simulations and full-custom layout. Thus the analog design process is very knowledge intensive. This situation leads to a poorer design time to silicon area ratio for analog designs and is one reason why digital solutions are often preferred. However, analog circuits are still required even in digital systems since most signals are analog by nature, and the analog-to-digital converter (ADC) and the digital-to-analog converter (DAC) circuits are partly analog.

During my first year at the department[1], I was working on a design tool for layout of analog integrated circuits. The research presented in this book concerns the other end of the design procedure, namely circuit analysis and synthesis. Computer aided symbolic circuit analysis is a blend of several different disciplines: applied electronics, circuit theory, control engineering, applied mathematics, numerical analysis and compiler theory.

---

1. Circuit Design Laboratory, Department of Applied Electronics, Lund Institute of Technology, Sweden.

Attention was given to symbolic analysis in the late 1960s and early 1970s and some tools were implemented to verify the analysis techniques (e.g. NAPPE [2]). The interest in this field of research was temporarily discouraged by the arrival of numerical circuit simulator programs (e.g. SPICE [56]). The use of these circuit simulators in practical analog circuit design is very common nowadays. The disadvantage, however, with numerical simulators is the difficulty to obtain an understanding of a circuit due to the large number of interacting parameters. Numerical simulators can be used to verify that a circuit has the intended behaviour, but they do not give any interpretation of the results or any directions on how to improve circuit performance.

In the late 1980s a renewed interest was devoted to the development of tools for symbolic analysis and analog design automation and several programs have been the result of this effort (e.g. FLOWUP [64], SAPEC [47], ASAP [21], ISAAC [30], IDAC [19], SSCNAP [46] and CASCA [28]). These symbolic tools are based on graph theory methods [64], [22], semi-numerical methods [67], [23], [2], [61] and determinant calculation methods [30], [47], [28]. Excellent surveys of methods and applications of symbolic analysis have been presented by Lin [48], [49].

Symbolic circuit analysis is a class of systematic methods (see chapter 3) that use mathematics to give an understanding of the electrical behaviour of a circuit. It does not replace neither circuit synthesis nor numerical circuit simulation, but instead it attempts to bridge the gap between them. Circuit analysis is sometimes confused with circuit design in much the same way as numerical simulations are often confused with analysis. Symbolic analysis allows us to improve circuit performance interactively, where the symbolic expressions give a mathematical interpretation of topological changes to a circuit. An analysis problem can often be formulated (either directly or indirectly) as the calculation of one (or several) transfer functions (e.g. gain, impedance, CMRR, pole-zero-locations, etc.). The modelling of the active circuit elements (i.e. transistors) is an important problem. While detailed general small-signal transistor models are suitable for numerical circuit analysis, simple or even ideal models are useful for synthesis, and transistor models of medium complexity are often appropriate for symbolic analysis. For the purpose of distortion analysis we have to replace the linearized transistor models with nonlinear models.

Two singular linear network elements: the nullator and the norator, were introduced by Carlin [9] in 1961, and the nullator-norator pair was given

1.1. Symbolic Analysis

the name nullor [10] in 1964 (see chapter 2). The usefulness of the nullor is due to its ability to model active networks independently of the actual realization of the active devices. The nullor has only had a minor impact on circuit design, which is unfortunate given its versatile nature. Carlosena and Moschytz have recently shown that current mode circuits, which are becoming more and more popular, can be derived from the ordinary voltage-based circuits by simply interchanging the nullators and the norators, and reversing the input and output nodes [11]. Further, they point out the convenience of using nodal analysis for these circuits.

The effects of the parasitics present in any realization of the active elements (e.g. transistors, operational amplifiers and current conveyors) may be taken into account by adding lumped elements to the nullor [55]. So, why is the nullor not present neither in the familiar hybrid-π bipolar transistor model nor in the Meyer MOS transistor model? The nullor is in fact implicitly present in both these transistor models, since a transconductance is equivalent to a nullor with emitter-degeneration (see section 7.3). We have studied transistor models suitable for circuit analysis and synthesis. The choice of appropriate models is essential to the circuit designer (see chapter 7).

The synthesis of negative-feedback amplifiers is an important part of analog circuit design. This book has been inspired by the work of Cherry and Hooper [15] in 1968 and a more systematic approach by Nordholt [58] in 1983. Our intuitive nullor synthesis method (see chapter 6) makes it possible to synthesize basic negative-feedback transistor amplifiers based only on simple equivalence rules for the nullor. We also show that familiar amplifier configurations (e.g. cascode, long-tailed pair, differential pair, two-stage transistor and compound transistor) can be obtained in a systematic way. With this method, basic transistor amplifier configurations no longer have to be considered as predefined, but instead they may evolve from the ideal nullor amplifier with its infinite gain. The argument that a synthesis method should always produce new amplifier configurations is sometimes encountered. However, the ability to assist the designer in locating a familiar amplifier circuit that is suitable for a given situation is to be considered as one of the major advantages of circuit synthesis.

Negative feedback reduces the influence of parameter variations as well as the nonlinear effects of the active elements. However, the amplifier may become unstable due to feedback. The location of the poles and zeros for the transfer function of the amplifier is therefore of interest for a circuit designer. A numerical method for the estimation of poles and zeros in linear circuits was presented by Haley and Hurst [36] in 1989. The approximate nature of the method, which apparently led to some controversy [34], [37], [38], is an essential part of our extended pole-splitting method that is used to obtain symbolic expressions for poles and zeros (see chapter 4). A first-order estimation is sufficient to approximate a dominant real pole but will fail in the presence of a dominant complex conjugated pole-pair. Haley and Hurst explain that in most cases second-order estimations are good approximations of the first two poles and zeros. We show that the poles (or zeros) to be splitted do not have to dominate at low frequencies, but it is sufficient if the they dominate at frequencies close to their locations (see sections 4.2 and 4.3). Thus, the high-order polynomials in the numerator and the denominator of the transfer functions can usually be factorized approximately as several low-order polynomials, which makes it much easier to find interpretable symbolic expressions for the poles and zeros. This becomes apparent when the transfer function is approximated at the frequency of a dominant pole (see section D.2).

The nodal analysis (NA) method (see chapter 5) which has been implemented in our symbolic analysis tool CASCA (see chapter 10) is often stated to be limited to $RLC - g_m$ networks and thus in some way inferior to modified nodal analysis (MNA). Branch equations describe the behaviour of circuit elements without a direct admittance representation in MNA [39]. However, the NA method is mathematically equivalent to the MNA method if gyrators (modelled with two transconductances) are used to introduce the necessary branch currents [29]. The NA method usually gives a very compact set of equations. Lin has presented a method for NA of networks containing nullors using the indefinite admittance matrix [49]. This method, compacted nodal analysis (CNA), results in an even more compact matrix than for the NA method. Compared to the number of equations required to describe the remaining network, the rank of the MNA matrix is increased by the number of nullors while the rank of the CNA matrix is reduced by the same number. The NA method also conveniently allows for the capacitive transsusceptance (i.e. transcapacitance or capacitive current generator), which is used in high-frequency MOS transistor models, and the inductive transsusceptance (i.e. inductive current generator). The usefulness of the compacted nodal analysis method cannot be overestimated. It has often been asserted that the NA method does not allow voltage-controlled voltage sources. However, we point

1.1. Symbolic Analysis

out that it is in fact possible to include voltage-controlled voltage sources even with arbitrary linear transfer functions in the nodal admittance matrix (see chapter 5). Our implementation of the NA method allows for the analysis of block diagrams, linear circuit schematics or combinations of the two. Further, we show that the transfer function $H(z)$ for a switched-capacitor network can be solved by using nodal analysis for our equivalent analog network (see chapter 9). Neither the ideal operational amplifier nor the switched capacitor are normally available for the NA method (the inductive transsusceptance is used to model the interaction between the clock phases).

So far only linear networks have been mentioned with the assumption that the nonlinear active elements (i.e. transistors) can be linearized at their operating points. However, this assumption is only valid in the presence of sufficiently small signal levels[1]. For example, since any given amplifier has a finite ability to deliver voltage and current the output signal will saturate if we try to force the amplifier beyond this limit, and some signal degradation (i.e. distortion) proportional to the magnitude of the signal will also occur below saturation. This degradation is often specified as a numerical value for total harmonic distortion (THD) given at some reference level, but the THD value does not give any interpretation of the nature of the distortion. The Volterra series method [42], [43], [57], [32] is useful for the analysis of soft nonlinearities in the presence of small signal levels but the method produces inaccurate results for hard nonlinearities. The describing function method [5], [18], [62] is an analytical form of the harmonic balance method [33], [41] and is applicable to soft as well as hard nonlinearities. We introduce this method in symbolic circuit analysis since it is conceptually interesting as an extended version of the frequency response method familiar from the analysis of linear networks. Describing function analysis of nonlinear systems is popular in control theory, where the method is usually (but not necessarily) restricted to a single nonlinear element. This means that if there is more than one nonlinear element, they must either be lumped together or all nonlinearities but the dominant one have to be neglected. For example, the behaviour of a bipolar (i.e. exponential) differential transistor pair is described by the function $\tanh(x)$, while the output stage of a feedback amplifier with reasonable gain is normally responsible for the major contribution to the total distortion. The fundamental assumption of the describing function method is that for a sinusoidal input signal only the fundamental frequency compo-

---

1. Signal refers to any carrier of information (e.g. voltage or current).

nent of the output signal has to be considered. This implies that there must be only an insignificant amount of higher-frequency harmonics present at the input of the nonlinearity (i.e. the feedback network should have low-pass properties). However, this assumption may be relaxed, which is necessary in order to calculate harmonic distortion. We use this relaxed version of the describing function method in combination with a general feedback model to obtain approximate symbolic expressions for the second- and third-order harmonic distortion components generated by a single transconductance nonlinearity embedded in an arbitrary linear network (see chapter 8).

1.2 Motivation

Let us consider the analysis of the common-mode rejection ratio (CMRR) for the CMOS differential amplifier in figure 1.1. The amplifier combines a differential gain stage and a differential-to-single-ended converter and is commonly used as input stage in two-stage operational amplifiers. The MOS transistors are modelled by a transconductance and an output conductance, see figure 1.2. It is assumed that $v_{BS} = 0$ for all devices, and hence no body effect occurs. The output conductance of the current source is denoted by g_o. A numerical analysis (e.g. in SPICE) would give $CMRR = 1 \cdot 10^4$, but does not indicate how this value can be improved. Hand calculations for this amplifier can be found in standard textbooks [35] and are repeated here for convenience. If it is assumed that g_{d3} for transistor Q_3 can be neglected since $g_{d3} \ll g_{m3}$, the differential gain for $g_{m3} = g_{m4} = g_{ml}$ and $g_{d3} = g_{d4} = g_{dl}$ is

$$A_{dm} = \frac{g_{mi}}{g_{dl} + g_{di}}, \qquad (1.1)$$

and the common-mode gain is

$$A_{cm} = -\frac{g_o g_{di}}{2 g_{ml}(g_{dl} + g_{di})}. \qquad (1.2)$$

Thus,

$$CMRR = \left| \frac{A_{dm}}{A_{cm}} \right| = 2 \frac{g_{mi} g_{ml}}{g_o g_{di}}. \qquad (1.3)$$

1.2. Motivation

If numerical values are inserted in (1.3), we get $CMRR = 2 \cdot 10^4$ which differs by a factor of 2 from the numerical value given earlier (?).

Now recalculate the expression for A_{cm} without neglecting the output conductance for transistor Q_3. At first, we do not make any assumptions about the load transistors being matched. The common-mode gain becomes,

$$A_{cm} = -\frac{g_o(g_{d3} + g_{di} + g_{m3} - g_{m4})}{g_{di}(g_{m3} + g_{m4}) + 2 g_{m3} g_{d4}}. \tag{1.4}$$

As we can see in the numerator of (1.4), if Q_3 and Q_4 are matched the assumption that $g_{d3} \ll g_{m3}$ cannot be used since g_{m3} is cancelled out by g_{m4} and we get

$$A_{cm} = -\frac{g_o}{2 g_{ml}}, \tag{1.5}$$

which gives

$$CMRR = 2\,\frac{g_{mi} g_{ml}}{g_o(g_{dl} + g_{di})}. \tag{1.6}$$

This example illustrates that circuit simplification based on topological knowledge has to be exercised with care. It is necessary to compare the approximated symbolic results with numerical simulations where no questionable assumptions are made. Calculations by hand are only feasible for small circuits and even then the calculations are tedious and the risk of introducing errors is high. However, the analytic results give a very good interpretation of the circuit behaviour.

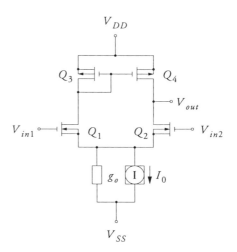

Figure 1.1: CMOS differential amplifier with active load.

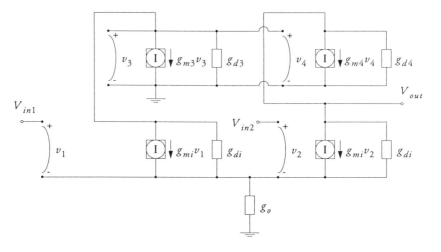

Figure 1.2: Small-signal equivalent circuit of the differential amplifier.

1.2. Motivation

Let us now consider a practical design example. Assume that we want to design an inverting voltage amplifier with unity gain to be used to convert an audio stereo power amplifier into a bridge-coupled mono power amplifier. We would like to achieve $THD < 0.02\%$ and an available voltage range of at least $V_{rms} = 2V$ is required. A first plausible idea is to use the transadmittance single-stage transistor amplifier where the output current is converted to voltage and inverted by the collector resistance R_C (figure 6.19). We know that the second-harmonic distortion HD_2 for a voltage-driven bipolar transistor varies linearly with the signal level (section 8.3.2) as

$$HD_2 = \frac{\hat{v}_{be}}{V_T} \frac{1}{4F}. \tag{1.7}$$

Thus, without feedback (i.e. $1/Y = R_E = 0$) and for the gain $A_0 = -200$ (i.e. $A_0 = -g_m R_C$) we get $HD_2 = 7\%$ at $v_o = 1.4V$. With appropriate feedback $R_E = R_C$ (i.e. $F = 1 + \beta A = 1 + g_m R_E$) this distortion is reduced to $HD_2 = 0.035\%$ at the same output level as before. Apparently this circuit generates more distortion than allowed by our specification and another problem is the poor output voltage range relative to the supply voltage

$$V_{CC} \frac{R_E}{R_E + R_C} = \frac{V_{CC}}{2} < V_o < V_{CC}. \tag{1.8}$$

In order to lower the distortion we may increase the gain A_0 and thereby also the return difference F (see section 8.1.1), or we may try to find a more linear nonlinearity. In chapter 6 the long-tailed pair transistor amplifier (figure 6.34) is suggested as a possible implementation of the nullor. This circuit allows for convenient negative feedback when cascaded with a common-emitter (CE) stage since it is non-inverting. Using our nullor synthesis method (chapter 6) it is easy to verify that the individual transistor stages (i.e. CE, CC and CB) as well as the cascades (i.e. CC-CB and CE-CC-CB) are all imperfect implementations of the nullor. From a small-signal point of view the long-tailed pair may be considered as a differential amplifier with single-ended output and the gain $g_m R_C / 2$. Alternatively the amplifier may be regarded as two cascaded CC-CB transistor stages with a relatively poor voltage coupling efficiency $\eta = 0.5$. It should be noted that neither the CC-stage nor the CB-stage invert the signal and that the CC-stage is required as a buffer following the CE-stage due to the low input impedance of the CB-stage. An important feature of the long-tailed pair is the nature of its nonlinear behaviour, see figure 1.3. Note that the often encountered assumption that the current generator may be considered as a short-circuit in order to simplify the calculation of the differential-mode gain (but not the common-mode gain) for the differential amplifier is only valid for small signal levels. It is clearly not valid in the case of a single-ended input or when the nonlinear effects are to be studied where the more general assumption that a current generator appears as an open-circuit to the signals is appropriate.

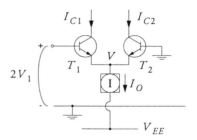

Figure 1.3: Long-tailed transistor pair.

The collector-current I_C for a bipolar transistor is exponential in the base-emitter voltage V_{BE} normalized to the thermal voltage V_T,

$$I_C = I_S e^{V_{BE}} . \qquad (1.9)$$

Applying this expression to the transistors T_1 and T_2 gives

$$\begin{cases} I_{C1} = I_S e^{2V_1 - V} \\ I_{C2} = I_S e^{0 - V} \end{cases} . \qquad (1.10)$$

The sum of the two collector currents must be equal to the current I_O,

$$I_O = I_{C1} + I_{C2} = I_S e^{-V}(e^{2V_1} + 1), \qquad (1.11)$$

1.2. Motivation

which gives the expression for the voltage V at the emitters,

$$e^{-V} = \frac{I_O}{I_S} \cdot \frac{1}{e^{2V_1}+1}. \tag{1.12}$$

Substituting eq. (1.12) into eqs. (1.10) we obtain the collector currents as

$$\begin{cases} I_{C1} = I_O \dfrac{e^{2V_1}}{e^{2V_1}+1} \\ I_{C2} = I_O \dfrac{1}{e^{2V_1}+1} \end{cases}. \tag{1.13}$$

The difference between the two currents becomes

$$I_{C1} - I_{C2} = I_O \frac{e^{2V_1}-1}{e^{2V_1}+1} \tag{1.14}$$

and can be rewritten as

$$I_{C1} - I_{C2} = I_O \frac{e^{V_1}-e^{-V_1}}{e^{V_1}+e^{-V_1}} = I_O \tanh(V_1). \tag{1.15}$$

The individual collector currents are

$$\begin{cases} I_{C1} = \dfrac{I_O}{2} + \dfrac{I_O}{2}\tanh(V_1) \\ I_{C2} = \dfrac{I_O}{2} - \dfrac{I_O}{2}\tanh(V_1) \end{cases}. \tag{1.16}$$

Thus, adding a long-tailed pair in cascade after the CE-stage gives us two advantages: the additional gain will reduce the signal level required at the input of the CE-stage and the significant part of the total distortion will thereby be generated by the output stage where $\tanh(x)$ is a more linear function than $\exp(x)$. The Taylor series expansion

$$\tanh(x) = x - \frac{1}{3}x^3 + \frac{2}{15}x^5 - \ldots \qquad (1.17)$$

indicates that this kind of nonlinearity will generate odd-harmonic distortion at its output. However, using the large-signal analysis methods presented in section 8.1.2 it can be shown that even a small bias voltage between the bases of the transistors T_1 and T_2 will produce second-harmonic distortion of a larger magnitude than that of the third-harmonic distortion for a perfectly matched transistor pair caused by the function $\tanh(x)$. The feedback configuration suggested for a voltage amplifier in section 6.2 is not able to provide an inverted output signal and we will therefore use the configuration in figure 1.4 instead, which is based on a transimpedance amplifier (figure 6.2) where the impedance Z_1 converts the input voltage to current. A possible realization of the suggested circuit is found in figure 1.5 where NPN transistors are used for the long-tailed transistor pair and a PNP transistor is used for the common-emitter input stage which makes it convenient to design the bias network. A test circuit with this configuration measured $THD = 0.005\%$ at $V_{in} = 1\,V_{rms}$ and has the output voltage range

$$V < V_o < V_{CC}. \qquad (1.18)$$

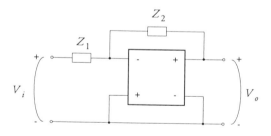

Figure 1.4: Inverting voltage amplifier.

$$A_V = \frac{V_o}{V_i} = -\frac{Z_2}{Z_1} \qquad (1.19)$$

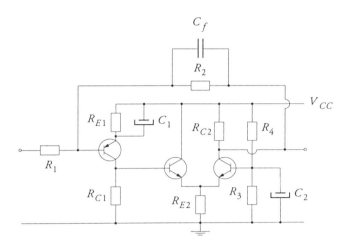

Figure 1.5: Inverting unity gain voltage amplifier.

1.3 Book Outline

Chapter 2 gives the reader basic knowledge about network theory and introduces some of the concepts found in the following chapters. Different formulation methods for linear circuit equations and some symbolic methods to solve these equations are presented in chapter 3. Chapter 4 presents the extended pole-splitting method which is used to obtain approximate symbolic expressions for the roots of a polynomial function. The compacted nodal analysis formulation is described in chapter 5. That the ideal nullor element can be used for transistor amplifier synthesis is shown in chapter 6. Chapter 7 presents different linear transistor models suitable for symbolic circuit analysis. In chapter 8, the describing function method is used for symbolic distortion analysis. The symbolic analysis of switched-capacitor networks using any tool intended for analysis of time-continuous networks is explained in chapter 9. Chapter 10 discusses a tool for symbolic analysis (i.e. CASCA) that has been implemented to study the ideas in this book. Chapter 11, finally, concludes the book.

Appendix A gives some basic determinant theory. The exact symbolic roots obtained with Maple for a cubic polynomial equation are given in appendix B. Appendix C contains the variable notation used for the describing function method in chapter 8. Some examples of circuit programs for CASCA are found in appendix D.

Chapter 2

Network Theory

The goal of this chapter is to give the reader basic knowledge in linear active network theory and to introduce some fundamental concepts. The nullor, which may be new to some readers, is presented.

2.1 Transmission Function

The behaviour of a linear network with the input signal $x(t)$ and the output signal $y(t)$ can be described by a linear nth-order differential equation with real coefficients a_i and b_i, and $n \geq m$,

$$a_0 \frac{d^n y}{dt^n} + a_1 \frac{d^{n-1} y}{dt^{n-1}} + \ldots + a_{n-1} \frac{dy}{dt} + a_n y = b_0 \frac{d^m x}{dt^m} + \ldots + b_m x. \qquad (2.1)$$

The Laplace transform [63] is defined as

$$L[f(t)](s) = \int_0^\infty f(t) e^{-st} dt. \qquad (2.2)$$

Let $X(s) = L[x(t)](s)$ and $Y(s) = L[y(t)](s)$ imply the Laplace transforms for the input and output signals respectively.

The Laplace transform of the time derivative is given by

$$L\left[\frac{d}{dt} f(t)\right] = sF(s) - f(0-), \qquad (2.3)$$

where $L[f(t)] = F(s)$, and $f(0-)$ is the value of $f(t)$ when $t = 0-$. Assuming zero initial state, i.e. the following initial conditions

$$\frac{d^{n-1}y}{dt^{n-1}}(0) = \ldots = y(0) = \frac{d^{m-1}x}{dt^{m-1}}(0) = \ldots = x(0) = 0, \quad (2.4)$$

eq. (2.1) can be rewritten as

$$(a_0 s^n + a_1 s^{n-1} + \ldots + a_{n-1} s + a_n) Y(s) = (b_0 s^m + \ldots + b_m) X(s). \quad (2.5)$$

The transmission function of a linear network is thus given as the ratio of the two polynomials in eq. (2.5)

$$H(s) = \frac{Y(s)}{X(s)} = \frac{b_0 s^m + \ldots + b_m}{a_0 s^n + a_1 s^{n-1} + \ldots + a_{n-1} s + a_n}. \quad (2.6)$$

Different methods available to determine the transmission function are treated in chapter 3. The major part of this book is devoted to time-continuous circuits but the same methods can also be used for time-discrete circuits, see chapter 9.

2.2 Poles and Zeros

The numerator and denominator polynomials in eq. (2.6) can be factorized. The numerator polynomial $N(s)$ has m roots called zeros, z_i, and the denominator polynomial $D(s)$ has n roots called poles, p_i. These roots can be either real or complex. The complex roots will always occur in complex conjugated pairs, since all the coefficients of both polynomials are real. A symbolic method to obtain approximate analytic expressions for the poles and zeros of eq. (2.7) is found in chapter 4.

$$H(s) = \frac{N(s)}{D(s)} = \frac{\sum_{i=0}^{m} b_i s^{m-i}}{\sum_{i=0}^{n} a_i s^{n-i}} = K \cdot \frac{\prod_{i=1}^{m}(s - z_i)}{\prod_{i=1}^{n}(s - p_i)}. \quad (2.7)$$

2.3. Nodes and Ports

Eq. (2.7) can be rewritten as

$$H(s) = A_0 \cdot \frac{\prod_{i=1}^{m}(1 - s/z_i)}{\prod_{i=1}^{n}(1 - s/p_i)} \tag{2.8}$$

where the low-frequency gain A_0 is given by

$$A_0 = H(0) = K \cdot \frac{\prod_{i=1}^{m}(-z_i)}{\prod_{i=1}^{n}(-p_i)}. \tag{2.9}$$

2.3 Nodes and Ports

Network elements are connected together at their terminals and a node is a connection point for one or several terminals. Current is defined to be positive when it is flowing away from a node and voltage can be measured at any node with respect to a reference node, usually the ground node. A pair of nodes, with a positive node and a negative reference node, is called a port. A two-terminal component (i.e. one-port) is called a branch. Examples of fundamental two-port networks are the controlled sources, the transistor and the operational amplifier. We obtain a one-port if the output port of the two-port is equal to the input port. Normally, a one-port is defined by one equation and a two-port by two equations[1]. The two-port equations are described in more detail in section 6.1.

---

1. This rule does not hold for singular network elements, i.e. the nullator and the norator.

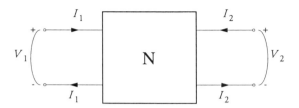

Figure 2.1: Two-port network.

2.4 Network Elements

This section introduces the basic passive circuit elements, the controlled sources and the gyrator. The nullor and some of its applications are also described.

2.4.1 Passive Elements

Resistors, capacitors and inductors are basic passive one-port network elements with admittance properties. They can be described in the time domain or in the complex frequency domain in terms of the Laplace transform operator s. The characteristics of these elements, i.e. the relation between the voltage across them and the current through them, are defined by their constitutive equations, see table 2.1.

Table 2.1: Constitutive equations for passive elements.

Y	Time-domain	Laplace-domain
G	$V = RI = \frac{1}{G}I$	$Y_R = \frac{I}{V} = G$
C	$I = \frac{dQ}{dt} = \frac{d(CV)}{dt} = C\frac{dV}{dt}$	$Y_C = \frac{I}{V} = sC$
L	$V = \frac{d\phi}{dt} = \frac{d(LI)}{dt} = L\frac{dI}{dt}$	$Y_L = \frac{I}{V} = \frac{1}{sL}$

2.4. Network Elements

2.4.2 Controlled Sources

A controlled source is a two-port network consisting of a current or voltage source whose value at the controlled output port depends on the current or voltage at the controlling input port. These controlled sources are ideal if they have either zero or infinite input impedance and either zero or infinite output impedance. Hence, there are four different controlled sources, the voltage-controlled current source, the voltage-controlled voltage source, the current-controlled voltage source and the current-controlled current source. The controlled sources are characterized by their transconductance g, voltage gain A_V, transresistance r and current gain A_I respectively.

2.4.3 Nonlinear Elements

Network models are necessary when the small-signal behaviour (i.e. AC analysis) of a linear network is to be calculated. Each component in the circuit schematic is replaced by its small-signal equivalent model. Transistors and diodes are examples of nonlinear elements whose parameters can be linearized at an operating point. We need to be able to model transistors and operational amplifiers as both ideal and nonideal devices. Ideal models may be used to study the behaviour of a circuit disregarding the parasitic effects involved. Linear transistor models that are suitable for symbolic analysis are discussed in chapter 7 and distortion analysis of nonlinear elements using describing functions is described in chapter 8.

2.4.4 Nullor

The nullor, see figure 2.2, consists of a nullator-norator pair [9]. These elements are called singular network elements [10]. They can be used to analyse network problems since ideal active elements (e.g. transistors and operational amplifiers) can be modelled with them. Nullors are described in, for example, [49], [54] and [55]. The inclusion of the nullor in nodal analysis is described in chapter 5 and the nullor is used to realize transistor amplifiers in chapter 6. The nullator, see figure 2.3 (a), is a port which is a short-circuit, $V = 0$, and at the same time it is an open-circuit, $I = 0$. The norator, see figure 2.3 (b), is a port with arbitrary voltage and current across it, i.e. V and I get the values that satisfy Kirchhoff's laws for the surrounding network. A non-singular one-port is defined by one equation and a two-port is defined by two equations. The nullator contributes two equations (one too many) and the norator contributes none (one too few), but the nullor combination has the required two equations. Thus the nullator and the norator always have to occur in pairs. It can be noted that a nullator in parallel with a norator is the equivalent of a short-circuit since $V = 0$ and I is arbitrary. In a similar way a nullator in series with a norator is the same as an open-circuit.

Figure 2.2: Nullor.

Figure 2.3: (a) Nullator and (b) norator.

2.4.5 Gyrator

The ideal gyrator [65] in figure 2.4 is a two-port network that is defined by

$$\begin{cases} I_1 = g_1 V_2 \\ I_2 = -g_2 V_1 \end{cases}, \qquad (2.10)$$

where $g_1, g_2 > 0$ and they are not necessarily equal. An important property of the gyrator is that anything connected to one of the ports appears as its duality at the opposite port. For example, it is well known that a capacitor becomes an inductor,

$$Z_1 = \frac{sC_2}{g_1 g_2} = sL_1. \qquad (2.11)$$

Gyrator transformations are used in chapter 5 and several other gyrator applications can be found in [53], [54]. The gyrator may be realized with a nullor as in figure 2.5. It is interesting to note that if the nullor circuit of figure 2.6

2.4. Network Elements

is used in figure 2.5 we are left with the gyrator since the resistors cancel out [10].

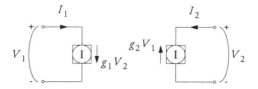

Figure 2.4: Ideal gyrator.

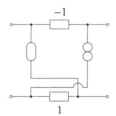

Figure 2.5: Equivalent circuit for a gyrator.

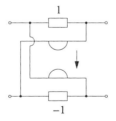

Figure 2.6: Nullor equivalent circuit.

Chapter 3

Symbolic Analysis

This chapter reviews different formulation methods for the circuit equations required to describe a network. The differences and similarities between symbolic and numerical analysis are discussed. An overview of some methods to obtain symbolic solutions for a set of linear circuit equations is given. The indefinite admittance matrix and its properties are described in detail.

Analytic analysis by hand based on Kirchhoff's current law (KCL) and Kirchhoff's voltage law (KVL) is the most frequently used method to calculate circuit behaviour in undergraduate electronics courses. However, in practical circuit design it is more common to use numerical circuit simulators. The disadvantage with such simulators is that it is difficult to obtain an understanding of a circuit due to the large number of parameters involved. These simulators can only be used to verify that a circuit has the intended behaviour. They do not give any directions on how to improve circuit performance. Hand calculations, on the other hand, are only applicable on smaller circuits. Still, analytic expressions are extremely useful for analog circuit design.

3.1   Circuit Equations

Any network obeys three basic laws: Kirchhoff's current law (KCL), Kirchhoff's voltage law (KVL) and the constitutive equations (CE) which sometimes are called branch relations (BR). Consider the small network in figure

3.1 with its oriented graph in figure 3.2. Kirchhoff's current law with currents flowing away from a node considered as positive gives

$$\begin{cases} -i_{b1} + i_{b4} + i_{b6} = 0 \\ -i_{b2} - i_{b4} + i_{b5} = 0 , \\ -i_{b3} - i_{b5} - i_{b6} = 0 \end{cases} \quad (3.1)$$

or, written in matrix form,

$$AI_b = 0, \quad (3.2)$$

where the matrix A is called the incidence matrix and I_b is the branch current vector. The matrix A has n rows equal to the number of ungrounded nodes and b columns equal to the number of branches.

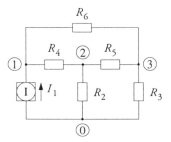

Figure 3.1: Small network.

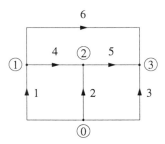

Figure 3.2: Oriented graph for the network.

3.1. Circuit Equations

Kirchhoff's voltage law gives

$$\begin{cases} v_{b1} = -v_{n1} \\ v_{b2} = -v_{n2} \\ v_{b3} = -v_{n3} \\ v_{b4} = v_{n1} - v_{n2} \\ v_{b5} = v_{n2} - v_{n3} \\ v_{b6} = v_{n1} - v_{n3} \end{cases}, \qquad (3.3)$$

or, written in matrix form,

$$V_b = A^T V_n, \qquad (3.4)$$

where V_b are the branch voltages and V_n are the node voltages with respect to the reference node. A^T is the transpose of the incidence matrix.

3.1.1 Tableau Formulation

The constitutive equations (CE) can be written as

$$y_1 v_b + k_1 i_b = i_1, \qquad (3.5)$$

or as

$$k_2 v_b + z_2 i_b = v_2, \qquad (3.6)$$

or more compactly in matrix form,

$$Y_b V_b + Z_b I_b = W_b. \qquad (3.7)$$

We get the tableau formulation by combining CE with KCL and KVL as

$$\begin{bmatrix} Z_b & Y_b & 0 \\ A & 0 & 0 \\ 0 & 1 & -A^T \end{bmatrix} \begin{bmatrix} I_b \\ V_b \\ V_n \end{bmatrix} = \begin{bmatrix} W_b \\ 0 \\ 0 \end{bmatrix}, \qquad (3.8)$$

or, more compactly,

$$TX = W. \qquad (3.9)$$

T is called the tableau matrix and is of rank $2b + n$.

3.1.2 Mesh Formulation

Either the mesh or the nodal formulation can be used to describe a circuit. The mesh equations are based on Kirchhoff's voltage law, which says that the sum of voltage drops around any closed loop is zero. The nodal equations are based on Kirchhoff's current law, which says that the algebraic sum of currents leaving any node is zero[1]. The required number of mesh equations is $b - n$ and the required number of nodal equations is n, where b is the number of branches in the network and n is the number of ungrounded nodes [8]. In general, at least two branches are connected to each node. This means that the number of mesh equations is typically greater than the number of nodal equations. Another important advantage of the nodal formulation is that it can be set up merely by inspection of the network. For the mesh equations, on other hand, we need to select a system of closed loops. Thus, setting up the mesh equations is normally much more difficult than setting up the nodal equations. This task becomes particularly difficult when the network is nonplanar, see figure 3.3, since a mesh is defined as planar.

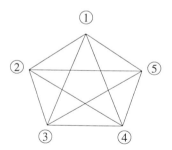

Figure 3.3: Nonplanar network.

---

1. A more general definition is: "In any cut which separates the network into two parts, the sum of currents in the cut edges is zero".

3.1. Circuit Equations

3.1.3 Loop Formulation

The limitation of mesh formulation (only by definition[1]) to planar networks is avoided if we use the more general loop formulation instead. With one node chosen as the reference node, we need six loop equations to describe the nonplanar network, see figure 3.3. When the nonplanar network is redrawn, as in figure 3.4, it becomes evident that the network is almost planar. With the branch 2-4 removed, we can easily find five of the required six loops or "windows". If we look at the network in three dimensions instead (it is nonplanar), see figure 3.5, a sixth "window" 2-4-3-2 that includes the branch 2-4 appears.

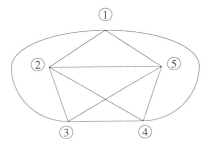

Figure 3.4: Almost planar network.

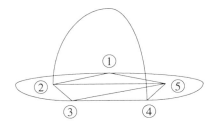

Figure 3.5: Three-dimensional view of the almost planar network.

---

1. It may be interesting to know that Bode [8] does not make any distinction between mesh and loop formulation. He also clearly states that nodal analysis is superior to mesh (loop) formulation.

3.1.4 Nodal Formulation

The tableau equations are

$$\begin{cases} Y_b V_b + Z_b I_b = W_b \\ AI_b = 0 \\ V_b = A^T V_n \end{cases} \qquad (3.10)$$

The last equation can be used to eliminate the branch voltages in the first equation, which gives

$$\begin{cases} Y_b A^T V_n + Z_b I_b = W_b \\ AI_b = 0 \end{cases} \qquad (3.11)$$

If we assume that all elements can be described as admittances and that only current sources are used, we get

$$\begin{cases} Y_b A^T V_n - I_b = -J_b \\ AI_b = 0 \end{cases} \qquad (3.12)$$

The first equation can be used to eliminate the branch currents in the second equation, which gives

$$A(Y_b A^T V_n + J_b) = 0, \qquad (3.13)$$

which can be rewritten as

$$AY_b A^T V_n = -AJ_b, \qquad (3.14)$$

or shorter

$$Y_n V_n = J_n. \qquad (3.15)$$

This equation is the nodal admittance formulation, where

$$Y_n = AY_b A^T \qquad (3.16)$$

3.1. Circuit Equations

is the nodal admittance matrix and

$$J_n = -AJ_b \qquad (3.17)$$

is the nodal current source vector. The rank of Y_n is equal to the number of ungrounded nodes.

Applying KCL to the network in figure 3.1, with the assumption that node 0 is grounded (i.e. chosen as reference node), gives the nodal equations in matrix form as

$$\begin{bmatrix} G_4 + G_6 & -G_4 & -G_6 \\ -G_4 & G_2 + G_4 + G_5 & -G_5 \\ -G_6 & -G_5 & G_3 + G_5 + G_6 \end{bmatrix} \begin{bmatrix} V_1 \\ V_2 \\ V_3 \end{bmatrix} = \begin{bmatrix} I_1 \\ 0 \\ 0 \end{bmatrix}. \qquad (3.18)$$

3.1.5 Modified Nodal Formulation

The nodal formulation can be modified to allow elements without an admittance representation,

$$\begin{cases} Y_n V_n + A I_b = J_n \\ Y_b A^T V_n + Z_b I_b = E_b \end{cases}. \qquad (3.19)$$

The branch currents through these elements are introduced in the nodal equations and they are defined in the branch equations. Branches without admittance descriptions are for example short-circuits, the input branch for current-controlled sources and the output branch for voltage sources. The size of this set of equations is $n + k$, where n is the number of ungrounded nodes and k is the number of introduced branch currents.

3.1.6 Summary

The tableau formulation is the most general formulation method. The disadvantage is that the tableau matrix is large even for small networks. The tableau method is free from term cancellations only if each element is represented with a unique symbol (i.e. circuit parameter) which is not always the case (e.g. mismatch analysis). The modified nodal formulation is more compact but still a general description of a circuit where branch currents are introduced only for branches without an admittance description. An even more compact matrix is obtained with the nodal formulation, which is applicable if all branches can be described as admittances. This is not a severe restriction since MOS transistors can be described directly by only transconductances and transcapacitances. The nodal formulation is, in fact, also a general method since all circuits can be described using transadmittances, see chapter 5. It is also possible to extend the nodal formulation method to handle ideal elements (i.e. nullors) conveniently, which makes the nodal formulation even more compact. A symbol that appears at several positions in the matrix in the two latter formulation methods may make it necessary to calculate terms that cancel in the final expression. This, of course, will increase the time it takes to solve the set of equations.

3.2 Symbolic Methods

Symbolic analysis is a class of methods where we use mathematics to obtain analytical expressions for the behaviour of a network. This section describes briefly the methods most often used for the symbolic analysis of linear networks: graph based methods, numerical methods and determinant methods. The purpose of all these methods is to solve a set of linear equations (see section 3.1) that describe the network.

3.2.1 Signal-Flow Graph Method

The linear circuit equations for the signal-flow graph method are written in the form $X = AX + BU$. The equations are represented by a weighted directed graph called a signal-flow graph. The signal-flow graph method is described in detail in for example [49] and [54]. The nodes represent variables and the branches, which are unidirectional, represent functional dependencies between the variables. The value of a node is the sum of all incoming signals. According to Cramer's rule in eq. (A.12), we need to calculate the determinant and cofactors of the matrix $(1 - A)$ to solve for X. Mason's formula gives the transfer function as

$$T_{ij} = \frac{x_i}{x_j} = \frac{\sum_k P_k \Delta_k}{\Delta}, \qquad (3.20)$$

3.2. Symbolic Methods

where $\Delta = |1 - A|$ is the network determinant, P_k is the weight of the kth path from the source node x_i to the dependent (sink or intermediate) node x_j, and Δ_k is the cofactor of the kth path, i.e. the determinant of the part of the graph that does not have any common nodes with the kth path. The weight of a path is the product of all branch weights along the path. The network determinant is calculated as

$$\Delta = 1 - \sum_{i} L_i + \sum_{i,j} L'_i L'_j - \sum_{i,j,k} L''_i L''_j L''_k + \ldots, \quad (3.21)$$

where L_i is any loop weight, $L'_i L'_j$ is any second-order loop weight and $L''_i L''_j L''_k$ is any third-order loop weight. An nth-order loop weight is the product of any n loops that do not have any nodes or branches in common. An efficient algorithm is needed for the enumeration of all paths and loops in the signal-flow graph. This method is suited for obtaining fully symbolic expressions for small networks. Larger networks can be analysed with the signal-flow graph method by decomposing them into smaller networks [64].

3.2.2 Tree Enumeration Method

The tree enumeration method is also a graph-based method. It is often referred to as a topological method. This method is applicable to networks containing only passive elements and transconductances. The determinant of the nodal admittance matrix can be calculated by enumerating all directed trees in a single directed graph. This is described in detail in [13]. A directed weighted graph is associated with the indefinite nodal admittance matrix. There is an edge directed from node i to node j with weight $-y_{ij}$, in the graph, for each entry in the matrix for which $y_{ij} \neq 0$ and $i \neq j$. The determinant of the nodal admittance matrix is equal to the sum of the branch-weight products for all directed trees in the directed graph with respect to a reference node. Topological methods are used to obtain fully symbolic network functions for small networks.

3.2.3 Interpolation Method

Assume that each entry in the admittance matrix Y is in the form $as + b$ where a and b are numerical values. The determinant is given by a polynomial in s,

$$|Y| = |sA + B| = c_0 + c_1 s + c_2 s^2 + \ldots + c_n s^n = \sum_{i=0}^{n} c_i s^i = P(s). \quad (3.22)$$

We can find the coefficients for that polynomial if we know $P(s)$ for $n + 1$ numerical values of s, that is

$$P(s_j) = \sum_{i=0}^{n} c_i s_j^i = y_j. \quad (3.23)$$

Thus we have $n + 1$ unknown coefficients and $n + 1$ equations and the coefficients are obtained by solving the equation system

$$\begin{bmatrix} 1 & s_0 & s_0^2 & \ldots & s_0^n \\ 1 & s_1 & s_1^2 & \ldots & s_1^n \\ \cdot & \cdot & \cdot & & \cdot \\ 1 & s_n & s_n^2 & \ldots & s_n^n \end{bmatrix} \begin{bmatrix} c_0 \\ c_1 \\ \cdot \\ c_n \end{bmatrix} = \begin{bmatrix} y_0 \\ y_1 \\ \cdot \\ y_n \end{bmatrix}. \quad (3.24)$$

This method is discussed in [23]. It is shown in [67] that the best result is obtained if the values for s_j are chosen so that they are complex and uniformly spaced on the unit circle. The fast Fourier transform can be used to solve the equations if we choose n so that $n + 1 = 2^m$. This method can be extended to handle circuits where not only the complex frequency s is a symbol. Numerical methods are more suited for large networks than the topological methods, especially when the only symbol is the complex frequency variable.

3.2. Symbolic Methods

3.2.4 Parameter Extraction Method

The parameter extraction method [2] and [61] combines the topological methods with fast numerical methods. The idea is to extract the symbolic parameters from the nodal admittance matrix and then numerically calculate the determinants for matrices containing only real numbers and the complex frequency s. This method is suited for obtaining partially symbolic expressions when only a small number of circuit elements are represented by symbols.

3.2.5 Determinant Methods

All methods mentioned so far solve a set of linear equations,

$$Ax = b. \qquad (3.25)$$

A more direct approach would be to use Cramer's rule in eq. (A.12). Thus, we need to calculate minors of the matrix A. Two well-known methods for the calculation of the determinant of matrices with numerical entries are variable elimination and Laplace's expansion. Examples of elimination algorithms are LU factorization and Gauss' method. The determinant can also be calculated using the determinant definition in eq. (A.2) or Laplace's expansion in eq. (A.5) along an arbitrary row or column. The determinant evaluation will be faster if the sparsity of the matrix is considered. The minors in the Laplace expansion can be recursively calculated and they may be stored and reused since, especially, the low-order minors are likely to be used several times [31].

3.2.6 Summary

The graph based methods are appropriate for obtaining fully symbolic transfer functions for small circuits. This has been our main interest. The parameter extraction method is suitable for obtaining mixed symbolic-numerical transfer functions for circuits where only a small number of elements are represented by symbols. The numerical methods (i.e. the interpolation method) may be the only way to handle larger networks where only the complex frequency s is a symbol. The determinant methods exploit determinant theory to calculate the network transfer functions. The problem is reduced to the calculation of cofactors (i.e. signed minor determinants). The complexity of a fully symbolic analysis may be considered as either factorial (from the determinant definition) or exponential (from the actual circuit topology) with respect to the number of nodes of the circuit.

3.3 Admittance Matrix

The admittance matrix is the core of the different formulation methods and thus deserves a thorough description. The equations can be set up by inspection of the network. Kirchhoff's current law for a node gives

$$\sum_{k=1}^{b} i_k = 0, \qquad (3.26)$$

where b is the number of branches connected to that node and i_k is the current in branch k. Applying this for all nodes gives an equation system containing the matrix Y_n,

$$Y_n V_n = J_n. \qquad (3.27)$$

This matrix exists in an indefinite and a definite form. The indefinite admittance matrix becomes definite if node k is grounded (i.e. chosen as reference node) and the corresponding row and column are deleted from the matrix.

The indefinite admittance matrix has some interesting properties. The sum of the elements in any row or column is zero,

$$\sum_i y_{ij} = \sum_j y_{ij} = 0. \qquad (3.28)$$

Such a matrix is singular (i.e. its determinant is zero),

$$|Y| = 0, \qquad (3.29)$$

and all its first-order cofactors (signed minors) are equal

$$\underline{Y}_j^i = \underline{Y}_n^n. \qquad (3.30)$$

An empty row n indicates that node n is connected neither to a transadmittance output nor to an admittance. An empty column n indicates that node

n is connected neither to a transadmittance input nor to an admittance. A node n is unconnected if both the row n and the column n are empty. That row and column can then be deleted.

3.4 Network Functions

The most important network functions for the network in figure 3.6 can be expressed as ratios of first and second-order cofactors for the indefinite admittance matrix. The matrix becomes definite when one column and one row are deleted. Column j can be deleted if we choose node j as reference node, since this makes $V_j = 0$. Row n can be deleted since $I_n = -I_m$ due to the current source. The freedom of choice in selecting nodes j and n makes the indefinite admittance matrix convenient to use when the input and output signals are balanced. The transfer functions are obtained with Cramer's rule.

The first-order cofactor is defined as

$$\underline{Y}_j^i = (-1)^{i+j} |Y_j^i|, \qquad (3.31)$$

where $|Y_j^i|$ is the determinant for the matrix Y with row i and column j deleted.

The second-order cofactor is similarly defined as

$$\underline{Y}_{ij}^{mn} = (-1)^{m+n+i+j} |Y_{ij}^{mn}|, \qquad (3.32)$$

where $|Y_{ij}^{mn}|$ is the determinant for the matrix Y with rows m and n, and columns i and j deleted.

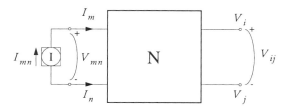

Figure 3.6: Two-port network.

The definite admittance matrix, with row n and column j deleted from Y, is denoted with A. Assume that we have the following set of nodal equations

$$\begin{bmatrix} a_{11} & a_{12} & \cdots & a_{1n} \\ a_{21} & a_{22} & \cdots & a_{2n} \\ \cdot & \cdot & \cdots & \cdot \\ \cdot & \cdot & \cdots & \cdot \\ a_{n1} & a_{n2} & \cdots & a_{nn} \end{bmatrix} \begin{bmatrix} V_1 \\ \cdot \\ \cdot \\ \cdot \\ V_n \end{bmatrix} = \begin{bmatrix} 0 \\ \cdot \\ I_m \\ \cdot \\ 0 \end{bmatrix}, \qquad (3.33)$$

and we want to solve for V_i. Cramer's rule in eq. (A.12) and Laplace's development in eq. (A.5) give

$$V_i = \frac{I_m \cdot \underline{A}_i^m}{|A|}, \qquad (3.34)$$

which can be expressed with cofactors of the Y matrix.

$$\begin{cases} I_{mn} = I_m = -I_n \\ V_{ij} = V_i - V_j = V_i \end{cases} \qquad (3.35)$$

give

$$Z_{mn}^{ij} = \frac{V_{ij}}{I_{mn}} = \frac{\underline{A}_i^m}{|A|} = \frac{(-1)^{m+i} |Y_{ij}^{mn}|}{|Y_j^n|} \cdot \frac{(-1)^{n+j}}{(-1)^{n+j}} = \frac{Y_{ij}^{mn}}{Y_j^n} = \frac{\underline{Y}_{ij}^{mn}}{\underline{Y}_n^n}. \quad (3.36)$$

This equation has to be modified in order to consider that the sign of the network functions depends on the polarity of the input and output ports, i.e. $V_{ij} = -V_{ji}$,

$$Z_{mn}^{ij} = \frac{V_{ij}}{I_{mn}} = \operatorname{sgn}(m-n)\operatorname{sgn}(i-j)\frac{\underline{Y}_{ij}^{mn}}{\underline{Y}_n^n}. \qquad (3.37)$$

3.5. Approximation Methods

The driving point impedance differs from the transfer impedance in that the input port and the output port are the same. It is used to calculate input and output impedance. Eq. (3.37) with $i = m$ and $j = n$ gives

$$Z_{mn} = \frac{V_{mn}}{I_{mn}} = \frac{Y_{mn}^{mn}}{Y_n^n}. \qquad (3.38)$$

The transfer voltage ratio is the ratio of the transfer impedance to the driving point impedance given in eqs. (3.37) and (3.38),

$$H_{mn}^{ij} = \frac{V_{ij}}{V_{mn}} = \operatorname{sgn}(m - n) \operatorname{sgn}(i - j) \frac{Y_{ij}^{mn}}{Y_{mn}^{mn}}. \qquad (3.39)$$

3.5 Approximation Methods

The exact analytic results produced by the different symbolic methods presented in this chapter are generally difficult to interpret due to the large number of terms involved. Fortunately, we are able to approximate these expressions based on the magnitude of the individual circuit parameters. There are several methods that can be used for the approximation procedure. CASCA truncates all terms that are some factor smaller than the largest term, and numerical values can be obtained for the remaining terms. This makes it possible to modify inappropriate numerical element values so that terms that the user wants to be insignificant are suppressed. Each coefficient in a polynomial of the complex frequency has to be treated individually in order to get an approximation that is valid at all frequencies. A polynomial can also be approximated at a fixed frequency. In this case terms from all coefficients are compared and the result is valid only in the vicinity of that frequency. Another method that gives a percentage value of the error introduced by the approximation is found in ISAAC [31]. All terms after the k most significant terms in an expression of the form

$$p(x) = \sum_{i=1}^{n} |a_i(x)| \qquad (3.40)$$

are neglected. The error is then given by

$$\varepsilon = \frac{\sum_{i=k+1}^{n} |a_i(x)|}{\sum_{i=1}^{n} |a_i(x)|} . \qquad (3.41)$$

This method may result in errors in the pole and zero locations, a problem that has been considered in ASAP [21], [22]. The ASAP program increases the error value ε gradually while monitoring the pole and zero displacements.

These methods suffer from the fact that all terms have to be calculated before the approximations can be carried out. For large networks, it would be preferable to be able to calculate only the dominant terms. Such a method could use numerical methods to decide which parts of a problem have to be evaluated symbolically. Some interesting research is in progress in this area [51], [68], [71].

Chapter 4

Extended Pole-Splitting

The application of negative feedback in amplifiers makes the overall gain less sensitive to parameter variations and reduces distortion. These improvements are affected by the same factor, namely the return difference[1]. The price we pay is that the overall gain is reduced and the amplifier may become unstable (i.e. oscillate) at some frequencies. The locations of poles and zeros of the transfer function give circuit designers insight into how feedback will affect performance and stability of the amplifier. Exact analytic expressions for the roots can only be obtained if the degree of the polynomial is four or less. However, it is possible to obtain approximate expressions if the roots are located far apart. The extended pole-splitting method [25], [27] is related to the numerical pole and zero estimation algorithm in [34], [36], [37], [38].

4.1 Numerical Roots

The calculation of approximate symbolic expressions for the roots requires that their numerical values are known, as will be shown in section 4.2. It is always possible to get numerical roots for a polynomial function,

$$f(s) = a_0 s^n + \ldots + a_{n-1} s + a_n = 0. \tag{4.1}$$

---

1. The return difference F is related to the loop gain T as $F=1+T=1+\beta A$.

We can use an iterative method, e.g. the Newton-Raphson method

$$x_{k+1} = x_k - \frac{f(x_k)}{f'(x_k)}, \qquad (4.2)$$

to obtain a first root. This root s_1 is then eliminated from $f(s)$,

$$a_0 s^n + a_1 s^{n-1} + \ldots + a_n = (b_0 s^{n-1} + \ldots + b_{n-1})(s - s_1). \qquad (4.3)$$

Identification of the coefficients gives Horner's scheme as

$$\begin{cases} b_0 = a_0 \\ b_1 = a_1 + b_0 s_1 \\ \ldots \\ b_{n-1} = a_{n-1} + b_{n-2} s_1 \end{cases} \qquad (4.4)$$

Assume that all real roots have been calculated and eliminated from $f(s)$. The remaining roots will then consist of complex conjugated pairs, since all the coefficients in the original polynomial are real. The Newton-Raphson method can be used to locate a complex root, if supplied with a complex starting value. We can avoid complex calculations by separating the real and imaginary parts of s, $f(s)$ and $f'(s)$. Assume that there is a complex conjugated root pair $s_{1,2} = u \pm iv$. We want to separate these roots from the polynomial

$$a_0 s^n + \ldots + a_n = (b_0 s^{n-2} + \ldots + b_{n-2})(s^2 - 2us + u^2 + v^2). \qquad (4.5)$$

Identification of the coefficients gives

$$\begin{cases} b_0 = a_0 \\ b_1 = a_1 + 2u b_0 \\ b_2 = a_2 + 2u b_1 - (u^2 + v^2) b_0 \\ \ldots \\ b_{n-2} = a_{n-2} + 2u b_{n-3} - (u^2 + v^2) b_{n-4} \end{cases} \qquad (4.6)$$

4.2 Approximate Symbolic Roots

Assume that we want to find the roots for a polynomial equation,

$$f(s) = a_0 s^n + \ldots + a_{n-1} s + a_n = 0. \tag{4.7}$$

At low frequencies this equation can be approximated with

$$a_{n-1} s + a_n = 0, \tag{4.8}$$

if there is a root

$$|s_1| \ll |s_2| \leq \ldots \leq |s_n|. \tag{4.9}$$

In this case the root is given by

$$s_1 = -a_n / a_{n-1}. \tag{4.10}$$

This root can now be eliminated from eq. (4.7). The roots for the remaining polynomial may be solved recursively. This method is called the pole-splitting technique. Analytic methods have been combined with the pole-splitting method in ASAP [21], [22]. The roots are approximated as long as the degree of the polynomial is larger than four, and thereafter the remaining roots can be solved with analytic methods. However, the pole-splitting method does not work if there are two closely located dominant roots, e.g. a complex conjugated pair, that is when

$$|s_1| \leq |s_2| \ll |s_3| \leq \ldots \leq |s_n|. \tag{4.11}$$

In this case eq. (4.10) is a poor approximation of only one of the dominant roots, but if we replace the first-order approximation in eq. (4.8) with a second-order approximation,

$$a_{n-2} s^2 + a_{n-1} s + a_n = 0, \tag{4.12}$$

we get a better estimate of the original polynomial at low frequencies. Approximate expressions for the polynomial coefficients in eq. (4.12) can then be used to obtain symbolic expressions for the two dominant roots. The numerical values supplied for the network parameters can be used to verify that the second-order approximation is valid, that is the numerical values for the two dominant roots for eq. (4.7) are compared with the roots for eq.

(4.12). In a similar way we can use third- or fourth-order approximations if necessary.

It is possible to extend the pole-splitting method even further. The pole or the poles to be splitted do not have to be dominant at low frequencies. It is sufficient if they are well separated from the other poles. That is, if we have a root s_k for which

$$|s_1| \leq \ldots \leq |s_{k-1}| \ll |s_k| \ll |s_{k+1}| \leq \ldots \leq |s_n|, \qquad (4.13)$$

we can approximate a polynomial

$$f(s) = a_0(s-s_1)\ldots(s-s_{k-1})(s-s_k)(s-s_{k+1})\ldots(s-s_n) \qquad (4.14)$$

close to the frequency $s = s_k$ as

$$f_1(s) = a_0 s^{k-1}(s-s_k)(-s_{k+1})\ldots(-s_n) = a_0 K s^k - a_0 K s_k s^{k-1} \qquad (4.15)$$

where

$$K = \prod_{i=k+1}^{n} (-s_i). \qquad (4.16)$$

The coefficients of the original polynomial in eqs. (4.7) and (4.14) are

$$a_\nu = (-1)^\nu a_0 \sum s_{i_1} s_{i_2} \ldots s_{i_\nu} \qquad (4.17)$$

for

$$1 \leq i_1 < i_2 < \ldots < i_\nu \leq n \qquad (4.18)$$

which gives

$$\begin{cases} a_{n-k} = (-1)^{n-k} a_0 \sum s_{i_1} s_{i_2} \ldots s_{i_{n-k}} \approx a_0 K \\ a_{n-k+1} = (-1)^{n-k+1} a_0 \sum s_{i_1} s_{i_2} \ldots s_{i_{n-k+1}} \approx -a_0 K s_k \end{cases} \qquad (4.19)$$

Thus the polynomial in eq. (4.14) is also approximated by

$$f_2(s) = a_{n-k} s^k + a_{n-k+1} s^{k-1} \qquad (4.20)$$

and the root becomes

$$s_k = -a_{n-k+1}/a_{n-k}. \qquad (4.21)$$

In a similar way, we can use higher order approximations as in eq. (4.12).

4.3 Two-Stage Operational Amplifier

The extended pole-splitting method is illustrated by the two-stage operational amplifier example in figure 4.1. We have used the Meyer model to expand the transistors. We may also introduce some simplifications to the small-signal schematic. The current generator (i.e. transistor Q_5) is assumed to be ideal and the influence of the C_{gb} capacitors are neglected. We can further exploit that $C_{gd6} \ll C_C$, $C_{gd7} \ll C_L$ and $C_{db7} \ll C_L$. The indices i and l are used for the input (i.e. Q_1 and Q_2) and load (i.e. Q_3 and Q_4) transistors respectively. A CASCA circuit program for the two-stage operational amplifier is found in appendix D.2.

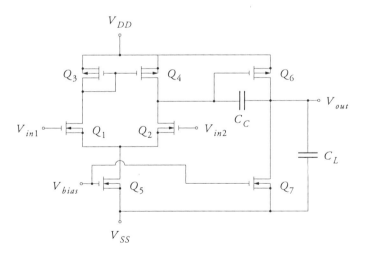

Figure 4.1: A two-stage operational amplifier.

The differential gain can be expressed as

$$H(s) = \frac{b_0 s^4 + b_1 s^3 + b_2 s^2 + b_3 s + b_4}{a_0 s^4 + a_1 s^3 + a_2 s^2 + a_3 s + a_4}, \tag{4.22}$$

and the approximated differential gain for low frequencies is

$$A_{dm} = \frac{b_4}{a_4} = \frac{g_{mi} g_{m6}}{(g_{di} + g_{dl})(g_{d6} + g_{d7})}. \tag{4.23}$$

The numerical values (in appendix D.2) for the poles indicate that they are well separated. Thus, the three first poles are given by

$$\begin{cases} p_1 = -\dfrac{a_4}{a_3} = -\dfrac{(g_{di} + g_{dl})(g_{d6} + g_{d7})}{g_{m6} C_C} \\ p_2 = -\dfrac{a_3}{a_2} = -\dfrac{g_{m6} C_C}{C_L (C_C + C_{gs6})} \\ p_3 = -\dfrac{a_2}{a_1} = -\dfrac{g_{ml}}{2 C_{gsl}} \end{cases} \tag{4.24}$$

The symbolic poles obtained for the three first-order approximations can be compared to those given by MAPLE [12] for the third-order approximation in appendix B. Clearly, the approximate symbolic poles in eqs. (4.24) are easier to interpret.

The numerical values for the zeros indicate that we have two closely located dominant zeros. Thus, they are given by

$$b_2 s^2 + b_3 s + b_4 = 0, \tag{4.25}$$

yielding

$$\begin{cases} z_1 = \dfrac{g_{m6}}{C_C} \\ z_2 = -\dfrac{g_{ml}}{C_{gsl}} \end{cases} \tag{4.26}$$

4.4. Single-Stage Operational Amplifier

Note that $2p_3 = z_2$. The locations of the poles and zeros before and after the symbolic approximations are found in figures 4.2 and 4.3. We see that first-order approximations of the two dominant zeros cause unacceptable displacements.

Figure 4.2: Displacements of (1) the original poles for (2) three first-order symbolic approximations given by eqs. (4.24).

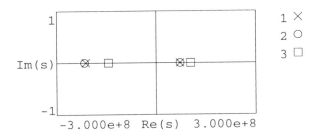

Figure 4.3: Displacements of (1) the original zeros for (2) one second-order symbolic approximation given by eqs. (4.26) and (3) two first-order approximations.

4.4 Single-Stage Operational Amplifier

Another example where the extended pole-splitting method can be used is the single-stage operational amplifier in figure 4.4. The indices i, l and o are used for the input (i.e. Q_1 and Q_2), load (i.e. Q_3 and Q_4) and output (i.e. Q_6 and Q_7) transistors respectively. Some of the capacitances have been lumped together,

$$\begin{cases} C_1 = C_{gs3} + C_{gs6} \\ C_2 = C_{gs4} + C_{gs7} \\ C_3 = C_{gs8} + C_{gs9} \\ C_4 = C_L + C_{db7} + C_{db8} \end{cases} \quad (4.27)$$

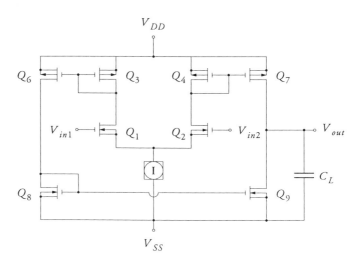

Figure 4.4: Single-stage operational amplifier.

The differential gain can be expressed as

$$H(s) = \frac{b_0 s^3 + b_1 s^2 + b_2 s + b_3}{a_0 s^4 + a_1 s^3 + a_2 s^2 + a_3 s + a_4}, \quad (4.28)$$

and the differential gain for low frequencies can be approximated as

$$A_{dm} = \frac{b_3}{a_4} = \frac{g_{mi}\, g_{mo}}{g_{ml}\, (g_{d7} + g_{d9})} \frac{g_{m8} + g_{m9}}{2 g_{m8}}. \quad (4.29)$$

The numerical values for the poles indicate that we have a dominant pole and a cluster of two poles. The first pole is then given by

$$p_1 = -\frac{a_4}{a_3} = -\frac{(g_{d7} + g_{d9})}{C_4}. \quad (4.30)$$

4.4. Single-Stage Operational Amplifier

The next two poles are given by

$$a_1 s^2 + a_2 s + a_3 = 0, \qquad (4.31)$$

as

$$\begin{cases} p_2 = -\dfrac{g_{m4}}{C_2} \\ p_3 = -\dfrac{g_{m3}}{C_1} \end{cases}, \qquad (4.32)$$

and the fourth pole is

$$p_4 = -\dfrac{a_1}{a_0} = -\dfrac{g_{m8}}{C_3}. \qquad (4.33)$$

The three zeros are well separated and can be obtained with first-order approximations as

$$\begin{cases} z_1 = -\dfrac{b_3}{b_2} = -\dfrac{g_{m1}}{C_{gs1} + C_{gso}} \\ z_2 = -\dfrac{b_2}{b_1} = -\dfrac{g_{m8} + g_{m9}}{C_3} \\ z_3 = -\dfrac{b_1}{b_0} = \dfrac{g_{m7}}{C_{gd7}} \end{cases}. \qquad (4.34)$$

Note that $p_2 = p_3 = z_1$ and $2p_4 = z_2$ if the transistors are matched. The locations of the poles and zeros before and after the symbolic approximations are found in figures 4.5 and 4.6. The fourth pole and the third zero are located outside of the graphs.

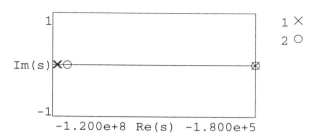

Figure 4.5: Displacements of (1) the original poles for (2) one first-order and one second-order symbolic approximation given by eqs. (4.30) and (4.32).

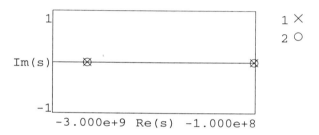

Figure 4.6: Displacements of (1) the original zeros for (2) two first-order symbolic approximations given by eqs. (4.34).

Chapter 5

Compacted Nodal Analysis

This chapter shows that the nodal formulation is a general method suitable for symbolic circuit analysis. Compacted nodal analysis results in matrix sizes that are even smaller than for nodal analysis. Voltage-controlled voltage sources with an arbitrary linear transfer function may be included in the compacted nodal admittance matrix.

Nodal analysis (NA) can be used for any network that can be modelled with transadmittances. The transadmittances may be conductive, capacitive or inductive current generators [25]. The modified nodal analysis [39] (MNA) equations may be obtained using nodal analysis and gyrator transformations [29]. Thus, we are not limited to $RLC - g_m$ networks. The nullor increases the rank of the modified nodal admittance matrix by one. The compacted nodal analysis (CNA), on the other hand, includes the nullor element as a matrix transformation, which reduces the rank by one instead [49].

5.1 Transadmittances

The nodal analysis method is conceptually interesting since we only need a single type of element, the transadmittance, see figure 5.1. This means that all "stamps" in the indefinite admittance matrix Y will appear as in table 5.1.

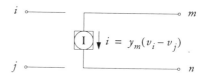

Figure 5.1: The transadmittance element.

The transadmittance may be a transconductance, a transcapacitance or an inductive transsusceptance, that is

$$\begin{cases} y_m = g \\ y_m = sC \\ y_m = 1/(sL) \end{cases} \qquad (5.1)$$

Table 5.1: Transadmittance element "stamp" in the indefinite admittance matrix.

Y	i	j
m	y_m	$-y_m$
n	$-y_m$	y_m

A common objection against nodal analysis is its alleged limitation to $RLC - g_m$ networks. The gyrator, which is modelled with two transconductances, can be used to model network elements which do not have direct admittance descriptions in the nodal admittance matrix, for example nullors (see figure 5.2), ideal transformers and voltage-controlled voltage sources [29]. Another objection often found is that a floating resistance generates four entries in the definite admittance matrix, which will give term cancellations during the determinant evaluation even if that resistor occurs only once in the network. The gyrator can be used to overcome this problem as well, see figure 5.3. Three of the four entries for G in the indefinite admittance matrix will now appear in the row or column corresponding to the reference node (i.e. ground).

5.2. Nullor

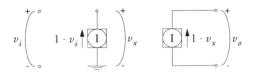

Figure 5.2: Nullor realized using two transconductances.

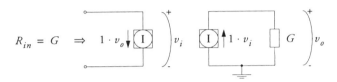

Figure 5.3: Floating resistance realized with the gyrator and a grounded conductance.

5.2 Nullor

The nullor is an ideal active element with infinite gain. Lets assume that we have added a nullator between the nodes i and j, and a norator between the nodes k and m. The current through the norator is denoted $I_{norator}$, and the nodal equations become

$$\begin{cases} y_{11}V_1 + \ldots + y_{1i}V_i + \ldots + y_{1j}V_j + \ldots + y_{1n}V_n = I_1 \\ \quad \vdots \\ y_{k1}V_1 + \ldots + y_{ki}V_i + \ldots + y_{kj}V_j + \ldots + y_{kn}V_n = I_k + I_{norator} \\ \quad \vdots \\ y_{m1}V_1 + \ldots + y_{mi}V_i + \ldots + y_{mj}V_j + \ldots + y_{mn}V_n = I_m - I_{norator} \\ \quad \vdots \\ y_{n1}V_1 + \ldots + y_{ni}V_i + \ldots + y_{nj}V_j + \ldots + y_{nn}V_n = I_n \end{cases} \quad (5.2)$$

We can apply $V_i = V_j = V_{i,j}$ and eliminate $I_{norator}$ by contracting the columns i and j, and the rows k and m. The new column and row are labelled (i, j) and (k, m), respectively. The compacted nodal equations then become

$$\begin{cases} y_{11}V_1 + \ldots + (y_{1i} + y_{1j})V_{i,j} + \ldots + y_{1n}V_n = I_1 \\ \vdots \\ (y_{k1} + y_{m1})V_1 + \ldots + (y_{ki} + y_{kj} + y_{mi} + y_{mj})V_{i,j} + \ldots = I_k + I_m. \\ \vdots \\ y_{n1}V_1 + \ldots + (y_{ni} + y_{nj})V_{i,j} + \ldots + y_{nn}V_n = I_n \end{cases} \quad (5.3)$$

5.3 Network Functions Using CNA

The network functions are given as the ratio of two cofactors. The calculation of a determinant and its sign factor is somewhat more complicated due to the row and column contractions caused by the nullor elements.

The first-order cofactor is defined as

$$\underline{Y}_{j_c}^{i_r} = (-1)^{i_r + j_c} \left| Y_{j_c}^{i_r} \right|, \quad (5.4)$$

where $\left| Y_{j_c}^{i_r} \right|$ is the determinant of matrix Y_C with the row labelled i and the column labelled j deleted.

The second-order cofactor is defined similarly as

$$\underline{Y}_{i_c j_c}^{m_r n_r} = (-1)^{m_r + n_r + i_c + j_c} \left| Y_{i_c j_c}^{m_r n_r} \right|, \quad (5.5)$$

where $\left| Y_{i_c j_c}^{m_r n_r} \right|$ is the determinant of matrix Y_C with the rows labelled m and n, and the columns labelled i and j deleted.

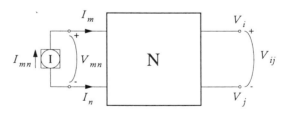

Figure 5.4: Two-port network.

5.4. Block Diagram Elements

The transfer impedance of the two-port network in figure 5.4 is given by

$$Z_{mn}^{ij} = \frac{V_{ij}}{I_{mn}} = \text{sgn}(m_r - n_r)\,\text{sgn}(i_c - j_c)\,\frac{Y_{i_c j_c}^{m_r,n_r}}{Y_{n_c}^{n_r}}, \qquad (5.6)$$

where we have considered the fact that the sign of the network function depends on the polarity of the input and output ports, i.e. $V_{ij} = -V_{ji}$. Eq. (5.6) with $i = m$ and $j = n$ gives the driving point impedance as

$$Z_{mn} = \frac{V_{mn}}{I_{mn}} = \text{sgn}(m_r - n_r)\,\text{sgn}(m_c - n_c)\,\frac{Y_{m_c n_c}^{m_r,n_r}}{Y_{n_c}^{n_r}}. \qquad (5.7)$$

The transfer voltage ratio is the ratio of the transfer impedance to the driving point impedance given in eqs. (5.6) and (5.7),

$$\begin{aligned}H_{mn}^{ij} &= \frac{V_{ij}}{V_{mn}} = \text{sgn}(m_c - n_c)\,\text{sgn}(i_c - j_c)\,\frac{Y_{i_c j_c}^{m_r,n_r}}{Y_{m_c n_c}^{m_r,n_r}} = \\ &= \text{sgn}(m_c - n_c)\,\text{sgn}(i_c - j_c)\,(-1)^{m_c + n_c + i_c + j_c}\,\frac{\left|Y_{i_c j_c}^{m_r,n_r}\right|}{\left|Y_{m_c n_c}^{m_r,n_r}\right|}.\end{aligned} \qquad (5.8)$$

5.4 Block Diagram Elements

We can describe an arbitrary voltage amplifier block, with finite gain and a number of poles and zeros, in the nodal admittance matrix. The transfer function may be written as the ratio of two polynomials. The voltage amplifier, in figure 5.5 (a), is described by

$$N(s)v_1 - D(s)v_o = 0, \qquad (5.9)$$

and the summation element, in figure 5.5 (b), is described by the equation

$$v_1 + v_2 - v_o = 0. \qquad (5.10)$$

Clearly, when all terms can be written in the form of yv_x, the equations (5.9) and (5.10) can be included in the set of nodal equations. However, we must define an interface for the block diagram in order to get a correct interaction between the block diagram and the surrounding network. Ideal voltage input and voltage output ports are obtained if we use nullor buffers to isolate the transfer function block, see figure 5.6. These block elements can be used to model, for example, the finite gain operational amplifier, as in figure 5.7.

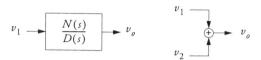

Figure 5.5: (a) Transfer function block and (b) summation element.

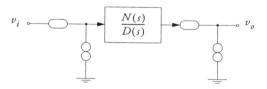

Figure 5.6: Buffered transfer function block.

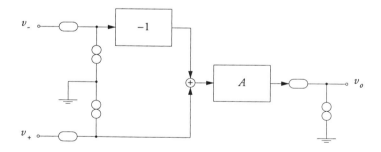

Figure 5.7: Block diagram for a finite gain operational amplifier.

Chapter 6

Nullor Synthesis

In this chapter we describe how the nullor element can be used to transform ideal feedback amplifiers into their corresponding transistor implementations. The nullor decouples the design of the feedback network from the active circuitry, and helps the designer to understand the synthesis of the four basic negative-feedback amplifiers. It is not our primary intention to create new circuits, but rather to obtain different amplifiers and to show how alternative amplifier configurations can be derived in a systematic way.

Electronics courses nowadays focus more on circuit analysis than on synthesis. If synthesis occasionally is included, it is in the form of reversed engineering. Existing or maybe somewhat modified circuits are analysed and used as building blocks. This may sometimes be practical but is not a very instructive approach. Basic transistor amplifier topologies are often said to be "well known" or "widely used" but no attempt is made to explain why they are configured the way they are. This is very unfortunate. However, it is possible to show that the transistors are in fact connected so that they will form imperfect nullor implementations.

The nullor is an ideal network element which is useful for circuit analysis but it can also be utilized for synthesis. Our objective is not to perform any detailed comparison of the properties of the different possible transistor configurations, but merely to illustrate the method. The fundamentals of feedback theory, the nullor concept and different transistor circuits have been available in papers and good textbooks for several decades [8], [55], [54], [11], [58], [14], [15], [10], [53]. However, it is still difficult in the literature to find a systematic and popular method that can be used for synthesis.

6.1 Two-Port Network

The two-port network in figure 6.1 has differential input and output ports to allow for different feedback loops (i.e. voltage input or current output). The currents are defined as positive into the network. The usual sign convention for the finite gain of the differential amplifier is used, which means that feedback between output and input terminals of opposite signs is negative.

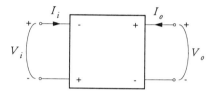

Figure 6.1: General two-port network.

There are six different sets of equations that will fully describe the relations between the input and the output variables of a two-port network. We will use the set called the transmission equations,

$$\begin{cases} V_i = a_{11}V_o + a_{12}I_o \\ I_i = a_{21}V_o + a_{22}I_o \end{cases}, \quad (6.1)$$

where the a-parameters, also called the chain matrix, are obtained as

$$\begin{aligned} a_{11} &= \left. \frac{V_i}{V_o} \right|_{I_o = 0} \\ a_{12} &= \left. \frac{V_i}{I_o} \right|_{V_o = 0} \\ a_{21} &= \left. \frac{I_i}{V_o} \right|_{I_o = 0} \\ a_{22} &= \left. \frac{I_i}{I_o} \right|_{V_o = 0} \end{aligned} \quad (6.2)$$

6.2. Basic Negative-Feedback Configurations

If we assume that negative feedback is applied to the two-port, the feedback gain is given by

$$A_f = \frac{A}{1+\beta A} = \frac{1}{\beta}\left(1 - \frac{1}{1+\beta A}\right) = \frac{1}{\beta}(1-\varepsilon). \tag{6.3}$$

The gain is $1/\beta$ for a feedback amplifier configuration where the active block is a nullor with infinite gain. The term ε takes into account the loss of gain due to finite loop gain for a transistor implementation. By assuming nullor characteristics for all amplifiers, the gain will be determined by the feedback networks only. When a suitable feedback network has been chosen (i.e. $1/\beta$), the number of transistor stages is determined by the loop gain required to make ε sufficiently small.

6.2 Basic Negative-Feedback Configurations

There are four basic single-loop negative-feedback configurations, see figures 6.2-6.5, that is the transimpedance, the transadmittance, the voltage, and the current amplifier. The input and output impedances for these amplifiers are either zero or infinite. The choice of feedback configuration is determined by the impedances of the source and the load. Our objective is to employ nullor transformations to find suitable transistor implementations for the embedded two-port.

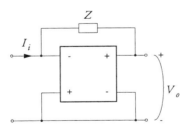

Figure 6.2: Transimpedance amplifier.

$$A_Z = \frac{V_o}{I_i} = -Z \tag{6.4}$$

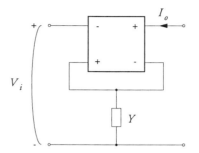

Figure 6.3: Transadmittance amplifier.

$$A_Y = \frac{I_o}{V_i} = Y \qquad (6.5)$$

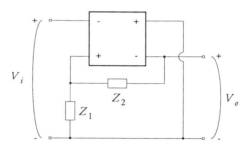

Figure 6.4: Voltage amplifier.

$$A_V = \frac{V_o}{V_i} = 1 + \frac{Z_2}{Z_1} \qquad (6.6)$$

6.3. Nullor Element

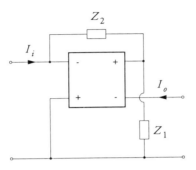

Figure 6.5: Current amplifier.

$$A_I = \frac{I_o}{I_i} = -\left(1 + \frac{Z_2}{Z_1}\right) \tag{6.7}$$

6.3 Nullor Element

For an ideal amplifier with infinite gain, all the a-parameters are equal to zero. Thus, eq. (6.1) can be rewritten as

$$\begin{cases} V_i = 0 \\ I_i = 0 \end{cases} \tag{6.8}$$

The output variables, V_o and I_o, take any values required by the feedback network to fulfil eq. (6.8). The input port is called a nullator and the output port is called a norator. The nullator and the norator must always appear as a pair, which is called a nullor, see figure 6.6. Note that the gain of the nullor has no defined sign[1] (i.e. it is infinite). The nullor can also be used to model the ideal transistor, as in figure 6.7. The behaviour of the ideal nullor transistor is given by eq. (6.8), which can be rewritten as

$$\begin{cases} V_E = V_B \\ I_E = I_C \end{cases} \tag{6.9}$$

---

1. This should not be confused with the signs for the two-port variables.

We can see that the voltage at the emitter will be equal to that at the base while the base current is zero (the terminology for the bipolar transistor is used). The ideal transistor will deliver the collector current necessary to accomplish this. Simple equivalence rules exist for the nullor [54], [55]. An open circuit between two nodes may be replaced by a nullator and a norator in series since $I = 0$ and V is arbitrary, see figure 6.8. A short circuit between two nodes may be replaced by a nullator and a norator in parallel since $V = 0$ and I is arbitrary, see figure 6.9. These equivalence rules will be used in the following to transform nullors to transistors and transistor pairs.

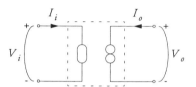

Figure 6.6: Nullor as an ideal two-port network.

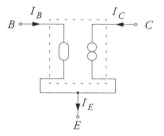

Figure 6.7: Nullor as an ideal transistor.

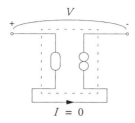

Figure 6.8: Nullor as an open circuit.

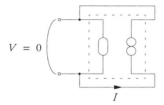

Figure 6.9: Nullor as a short circuit.

6.4 Nonideal Transistor

The nonideal transistor has four disadvantages compared to the ideal nullor transistor: parasitic impedances result in finite bandwidth, the gain is finite, the transistor requires a bias voltage, and it shows nonlinear behaviour. By assuming, without loss of generality, a bipolar transistor and neglecting the parasitics we get the transistor model in figure 6.10.

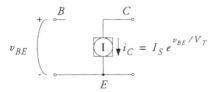

Figure 6.10: Bipolar transistor with nonlinear and finite gain.

The output current is given by

$$i_C(v_{BE}) = I_S e^{v_{BE}/V_T} = I_S e^{(V_{BEQ}+v_{be})/V_T} = I_{CQ} e^{v_{be}/V_T},\qquad(6.10)$$

where V_{BEQ} is the bias level, v_{be} is the signal component at the input and I_{CQ} is the quiescent current at the output. The gain becomes

$$A_Y = \frac{di_C}{dv_{BE}} = \frac{I_{CQ}}{V_T} e^{v_{be}/V_T},\qquad(6.11)$$

and for small input levels

$$A_Y = \left.\frac{di_C}{dv_{BE}}\right|_{v_{be}\to 0} = \frac{I_{CQ}}{V_T} = g_m.\qquad(6.12)$$

For low frequencies the nonideal transistor will behave almost like the ideal nullor transistor provided that a bias current I_{CQ} is supplied and that the input voltage v_{be} is kept small. The arguments may be repeated for other transistor types.

6.5   Transistor Implementations of the Nullor

Nonideal transistors can be used to form imperfect implementations of the nullor. A nullor implementation that can be used for all four single-loop feedback configurations requires two transistors, since the input and the output ports must have separated terminals. The single NPN transistor implementation in figure 6.11, which follows directly from figure 6.7, can only be used for the transimpedance and the transadmittance amplifiers. The PNP transistor (see figure 6.12) is easily recognized if the nullor transistor in figure 6.7 is flipped vertically. Note the presence of signs at the input and the output ports due to the finite gain (compare with figure 6.1). The two-port signs in figures 6.2-6.5 define the relative orientation of the gain and they can thus be reversed whenever it makes the translation to transistor implementations more convenient. Furthermore, we must supply a quiescent point for each transistor. An alternative nullor configuration is shown in figure 6.13. We have added an open-circuit nullor to the ordinary nullor to make it possible

6.5. Transistor Implementations of the Nullor

to identify two transistors. That is, each nullor must have a connection between its input and output as in figure 6.7 to be replaceable by a transistor. We can use any out of four different combinations of NPN and PNP transistors. The choice of transistors is influenced by their large-signal behaviour, that is we must consider the bias problem and the linearity of the nullor implementation. The configuration with separated input and output terminals in figure 6.14 is implemented with two NPN transistors. The configuration in figure 6.15 uses an NPN-PNP transistor pair instead. The latter circuit becomes similar to the single transistor implementation in figure 6.11 if we connect the negative output terminal to the positive input terminal, and is then called the compound transistor configuration. Yet another nullor configuration is the balanced transconductance with infinite gain shown in figure 6.16 (see also figures 7.4 and 9.1). Two nullators are connected in series at the input and two norators are connected in series at the output. We still have $V_i = I_i = 0$, while the output variables take arbitrary values. No current will flow through the added short circuit, since the current is zero into both input terminals. This configuration is implemented by the differential transistor pair shown in figure 6.17 where the emitters of the transistors correspond to the internal node of the balanced nullor.

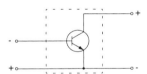

Figure 6.11: Nullor implemented by a single NPN transistor.

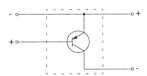

Figure 6.12: Nullor implemented by a single PNP transistor.

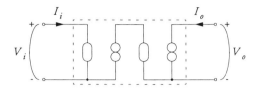

Figure 6.13: Alternative nullor configuration (two-stage transistor pair).

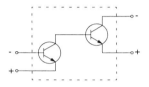

Figure 6.14: Nullor implemented by a two-stage NPN-NPN transistor pair.

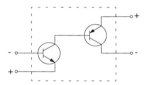

Figure 6.15: Nullor implemented by a two-stage NPN-PNP transistor pair.

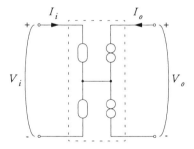

Figure 6.16: Balanced nullor configuration (differential transistor pair).

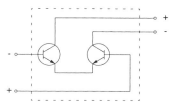

Figure 6.17: Nullor implemented by a differential transistor pair.

6.6 Transistor Amplifiers

In this section, we will illustrate our synthesis method with some examples. A few (i.e. four) different transistor implementations of the nullor are used to replace the two-port network in the basic negative-feedback configurations. Collector and emitter resistances are added for bias purposes when necessary. The transimpedance and transadmittance amplifiers can be implemented with a single transistor, see figures 6.18 and 6.19. This is obvious if we study the negative-feedback configurations (figures 6.2 and 6.3, respectively) and the nullor implementation in figure 6.11. The voltage and current amplifiers require a two-stage transistor pair configuration, see figures 6.20 and 6.21. This is a consequence of their feedback configurations (figures 6.4 and 6.5, respectively) and the nullor implementation in figure 6.14. Multi-stage feedback amplifiers can, in general, be derived by adding one or several open-circuit or short-circuit nullors, as in figure 6.13. The transimpedance and transadmittance amplifiers can also be implemented with an NPN-PNP transistor pair (figure 6.15), see figures 6.22 and 6.23. The single transistor is here replaced by a compound transistor configuration. The voltage and current amplifiers can also be realized with a differential transistor pair configuration, as in figures 6.24 and 6.25. This is apparent from the basic feedback configurations (figures 6.4 and 6.5, respectively) and the balanced nullor implementation in figure 6.17.

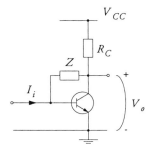

Figure 6.18: Transimpedance single-stage transistor amplifier.

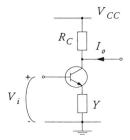

Figure 6.19: Transadmittance single-stage transistor amplifier.

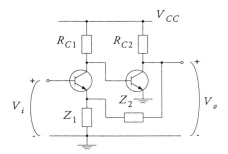

Figure 6.20: Voltage two-stage transistor amplifier.

6.6. Transistor Amplifiers

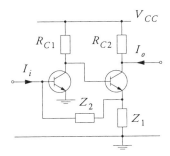

Figure 6.21: Current two-stage transistor amplifier.

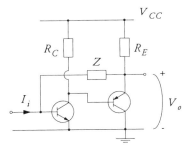

Figure 6.22: Transimpedance compound transistor amplifier.

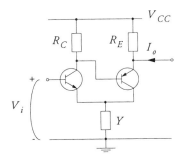

Figure 6.23: Transadmittance compound transistor amplifier.

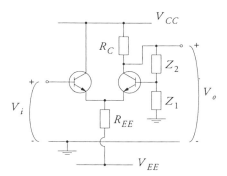

Figure 6.24: Voltage differential transistor amplifier.

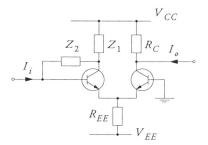

Figure 6.25: Current differential transistor amplifier.

6.7 Alternative Transistor Amplifiers

Some alternative transistor implementations of the nullor can be obtained by using the short circuit equivalence rule in figure 6.9. In figure 6.26 we replace a cut below the norator in the ordinary nullor transistor (figure 6.7) with a short-circuit nullor. The result is the CE-CB transistor configuration (cascode) in figure 6.27. In figure 6.28 we have replaced a cut at the emitter of the nullor transistor instead. This configuration is implemented by the CC-CB transistor cascade (long-tailed pair) in figure 6.29. In figure 6.30 we have replaced two cuts between the norators in figure 6.16 with short-circuit nullors and the result is the cascode transistor configuration shown in figure 6.31. The transimpedance (figure 6.2) and transadmittance (figure 6.3)

6.7. Alternative Transistor Amplifiers

amplifiers can be implemented with the cascode configuration (figure 6.27), see figures 6.32 and 6.33 respectively. Note that the base of the upper transistor in figure 6.33 is connected to the emitter of the lower transistor and not to signal ground since that would ruin the global feedback and reduce the circuit to a cascade of two local feedback stages: the transadmittance (figure 6.19) and the current follower (figure 6.36). The non-inverting long-tailed pair transistor amplifier in figure 6.34 has to be cascaded with an inverting amplifier in order to implement a nullor that allows for negative feedback. The CC and CB transistor stages which are special cases of the voltage (figure 6.4) and current (figure 6.5) amplifiers can also be derived independently, see figures 6.35 and 6.36, where the nullor is replaced by an NPN (figure 6.11) and a PNP (figure 6.12) transistor respectively. The cascode differential amplifier in figure 6.37 is an alternative to the standard differential amplifier configuration. Again, note that the bases of the upper transistors should be connected to the emitters of the lower transistors and not to signal ground.

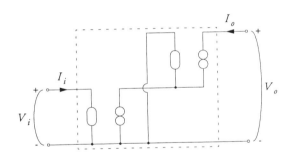

Figure 6.26: Nullor configuration (cascode).

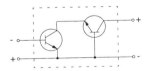

Figure 6.27: Nullor implemented by a CE-CB transistor cascade (cascode).

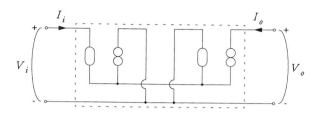

Figure 6.28: Alternative nullor configuration (long-tailed pair).

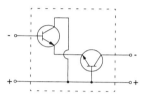

Figure 6.29: Nullor implemented by a CC-CB transistor cascade (long-tailed pair).

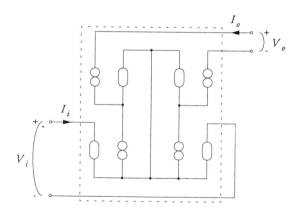

Figure 6.30: Cascode balanced nullor.

6.7. Alternative Transistor Amplifiers

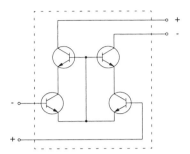

Figure 6.31: Nullor implemented by a cascode differential transistor configuration.

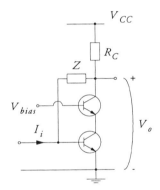

Figure 6.32: Transimpedance cascode transistor amplifier.

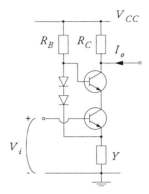

Figure 6.33: Transadmittance cascode transistor amplifier.

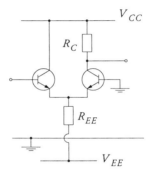

Figure 6.34: Long-tailed pair transistor amplifier.

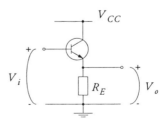

Figure 6.35: Voltage follower (CC stage).

6.7. Alternative Transistor Amplifiers

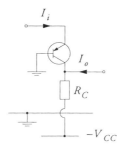

Figure 6.36: Current follower (CB stage).

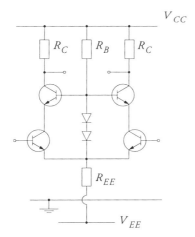

Figure 6.37: Cascode differential amplifier.

Chapter 7

Transistor Models

This chapter presents different linear transistor models suitable for computer aided symbolic circuit analysis. We discuss why a more refined model does not necessarily give a more useful symbolic result. It is also concluded that a simple, or even ideal, transistor model can have great advantages in circuit synthesis.

The analysis and design of integrated circuits depend on the application of suitable models for the active nonlinear circuit elements (i.e. transistors). While an ideal transistor model is appropriate for an initial synthesis approach a more refined transistor model may be required to analyse the actual realization of the circuit. We may also want to avoid having to model all transistors in a circuit at the same level of refinement. The nullor concept [54] has been found to be a very useful tool which can be used to understand why some of the basic transistor amplifier circuits are configured the way they are. Other methods are usually focused on analysis while the nullor approach is well suited for synthesis as well [20], [58], [11]. Once a proper configuration has been found the nonideal properties of the circuit need to be investigated. Characteristics of interest are for example the input and output impedances, and the voltage transfer ratio. A first approach to tackle this analysis may be to consider the transistor as linear for small signals. All transistor parameters can then be linearized at an operating point. One of the reasons why operational amplifiers are so popular is that their behaviour is close to that of the ideal nullor amplifier, at least for low frequencies. Another reason is the fact that they have separate terminals for bias voltages and input signals. Small-signal transistor models easily become so focused on parasitic phenomena that the concept of an ideal transistor is lost [55]. It is important to remember that there is always an ideal transistor embedded, even in the most refined models.

7.1 Network Elements

The fundamental two-port network elements used in this chapter are the transconductance, the transcapacitance and the nullor, see figure 7.1. The choice of elements is based on the fact that the MOS transistor as well as the bipolar transistor can be modelled conveniently by combinations of voltage-controlled current sources. The nullor, which is described in more detail in section 7.3, is used to model ideal transistors. All three elements are well suited for nodal analysis [25].

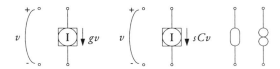

Figure 7.1: Fundamental network elements.

7.2 Nodal Formulation

Nodal formulation can be used if we assume that all network elements can be described with admittances and current sources. This is a reasonable assumption for transistor circuits. An important property of the nodal formulation is that it can be set up merely by inspection of the network. This makes the method well suited for hand calculations as well as computer analysis. Kirchhoff's current law is applied at each node in a circuit to obtain

$$Y_n V_n = J_n, \qquad (7.1)$$

where Y_n is the nodal admittance matrix, V_n is the nodal voltage vector, and J_n is the nodal current source vector. Nullors cannot be described conveniently in the admittance matrix. Instead, we may consider the nullors as additional conditions imposed on the nodal formulation. The two columns corresponding to the input nodes of a nullor can be added since the two input node voltages are equal, and the two rows corresponding to the output nodes are added to eliminate the output current [49]. Thus, each nullor reduces the rank of the matrix by one (see section 5.2).

7.3 Nullor

The nullor is an extremely useful network element [20], [54], [58], [11]. It can be used to model generic amplifiers, as in figure 7.2, and generic transistors, as in figure 7.3, without introducing any parasitic effects (i.e. symbolic parameters). The nullator, at the input, is a port which is a short-circuit, $V_i = 0$, and at the same time it is an open-circuit, $I_i = 0$. The norator, at the output, is a port with arbitrary voltage and current across it, that is V_o and I_o take the values that satisfy Kirchhoff's laws for the surrounding network (i.e. the conditions set by the nullator). In figure 7.4, we see that a transconductance is equal to an ideal transistor with local feedback (i.e. emitter-degeneration).

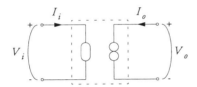

Figure 7.2: Nullor as a generic amplifier.

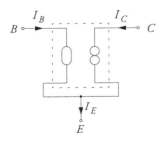

Figure 7.3: Nullor as a generic transistor.

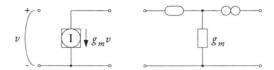

Figure 7.4: (a) Transconductance (b) using a nullor.

7.4 Small-Signal Transistor Models

Small-signal transistor models are required to obtain network transfer functions, for example the voltage transfer ratio, for a transistor circuit. The nonlinear behaviour of a transistor is then linearized at an operating point. The linearization is only valid under the assumption of soft nonlinearities in combination with small signal levels. Symbolic analysis imposes somewhat different requirements on the transistor models than numerical analysis does. A numerical result becomes more detailed with a refined transistor model but the complexity of the result is the same. A symbolic result, on the other hand, becomes much more complicated with the inclusion of parasitic effects. It can therefore sometimes be preferable to use a less detailed model. The difference between a basic MOS transistor model and a high-frequency model is shown in figures 7.5 and 7.6. The basic model considers the transconductance, the output conductance and the input capacitance. The high-frequency model is charge-oriented and includes transcapacitances to model the nonreciprocal capacitance behaviour [50], [66], [69]. Also included are a number of parasitic capacitances. It is possible to find medium-frequency models that are a compromise between these two extremes. The popular Meyer model [52], for example, is obtained if we neglect the transcapacitances. The bipolar transistor model in figure 7.7 is similar to the basic MOS transistor model. The only difference is that the MOS transistor has a purely capacitive input. The hybrid-π bipolar transistor model in figure 7.8 includes a number of parasitic effects, of which the resistances in series with the transistor terminals are the most unfortunate since they result in three additional internal circuit nodes. From a symbolic analysis point of view this is a severe problem since the complexity of the purely symbolic expressions is $O(n!)$ where n is the number of circuit nodes.

7.4. Small-Signal Transistor Models

The choice of a suitable transistor model varies with the application. The Meyer model can be used to calculate open-loop gain, and poles and zeros for a two-stage MOS operational amplifier. If the same amplifier is used as an inverting amplifier, the low-frequency voltage gain is given by the external resistances, i.e. $A = -R_2/R_1$. Clearly, to obtain this result the choice of transistor model is of minor importance. A transistor that operates as a current source can be modelled with a simpler model (e.g. conductance or open-circuit). In general, less refined models are used at higher levels of hierarchy. Only a few symbolic tools seem to have realized the need for transistor models at different levels of complexity (e.g. [22] and [31]). It is the simplicity of its model that has led to the popularity of the operational amplifier. Therefore, the applicability of the less refined transistor models should not be underestimated.

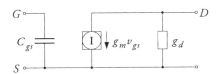

Figure 7.5: Basic MOS transistor model.

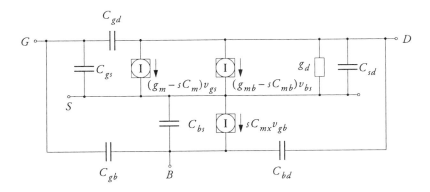

Figure 7.6: High-frequency MOS transistor model.

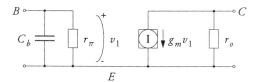

Figure 7.7: Basic bipolar transistor model.

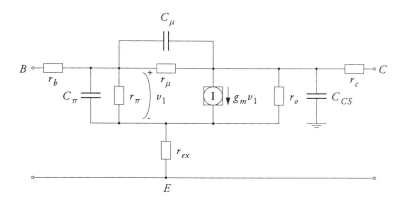

Figure 7.8: Hybrid-π bipolar transistor model.

7.5 Approximate Symbolic Expressions

It is evident from the determinant definition that the symbolic transfer functions contain a large number of terms. The indefinite admittance matrix may be sparse, but on the other hand each entry often contains more than one symbol. Even for a small circuit, such as the single-stage bipolar transistor amplifier in figure 7.9, we will get many terms. The transfer voltage ratio has 23 terms in the numerator and 165 terms in the denominator, if we use the bipolar hybrid-π transistor model shown in figure 7.8.

7.5. Approximate Symbolic Expressions

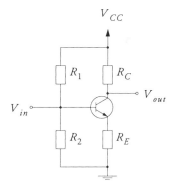

Figure 7.9: Single-stage bipolar transistor amplifier.

Thus, the exact symbolic transfer functions are difficult to interpret due to their complexity. All circuit elements are therefore assigned numerical values that are used to approximate the symbolic expressions. If all small terms in both the numerator and the denominator are removed we get the approximated transfer voltage ratio

$$A = \frac{V_{out}}{V_{in}} = -\frac{g_m R_C}{1 + g_m R_E}. \tag{7.2}$$

Assuming that the transconductance is sufficiently high (i.e. $g_m R_E \gg 1$) we get

$$A = \frac{V_{out}}{V_{in}} = -\frac{R_C}{R_E}, \tag{7.3}$$

which is to be expected if the transistor is nearly ideal. If eq. (7.2) has the required accuracy we could have used the basic transistor model instead. This would have given a transfer function with 3 terms in the numerator and 8 terms in the denominator. If eq. (7.3) is sufficient, the transistor may be modelled with a nullor. Furthermore, it can be noted that the exact expression we get when we use the basic bipolar model cannot be obtained by truncating the expression for the complete hybrid-π model, but it is of course possible to approximate both expression so that they become equal. This is quite obvious, and is best illustrated by the following small example. The exact voltage ratio transfer function at low frequencies for the amplifier, when the transistor is modelled with the basic transistor model, is

$$A = -\frac{g_m r_o r_\pi R_C - R_E R_C}{g_m r_o r_\pi R_E + r_o r_\pi + R_E r_o + R_C r_\pi + R_E r_\pi + R_C R_E}. \quad (7.4)$$

The terms in the numerator and the denominator are ordered by their magnitudes. If we use the hybrid-π transistor model instead, and the same number of terms are kept in the numerator and the denominator, the approximated result is

$$A = -\frac{g_m r_o r_\pi R_C - R_E R_C}{g_m r_o r_\pi R_E + r_o r_\pi + g_m r_o r_\pi r_{ex} + r_b r_o + R_E r_o + R_C r_\pi}. \quad (7.5)$$

If we compare the denominators of these two expressions, we find that only four of the six terms are the same and that the third and fourth largest terms in eq. (7.5) are not present in eq. (7.4).

When negative feedback is applied to the transistor, the feedback gain is given by

$$A_f = \frac{1}{\beta}(1-\varepsilon). \quad (7.6)$$

The gain is $1/\beta$ if the transistor is ideal. The term ε takes into account the parasitics in the real transistor. By assuming an ideal transistor, the gain will be determined by the feedback network only (i.e. $A_f = 1/\beta$) even if inappropriate (if a real transistor is used) numerical values are assigned to the resistors. Thus, a simple transistor model can in fact sometimes give a better understanding of the ideal behaviour of a circuit than a more refined model can do. Eq. (7.6) for eq. (7.5) becomes,

$$A_f = -\frac{R_C}{R_E}\left(1 - \frac{R_E^2 + r_b r_o + R_E r_o + R_C r_\pi + r_o r_\pi + g_m r_{ex} r_o r_\pi}{r_b r_o + R_E r_o + R_C r_\pi + r_o r_\pi + g_m R_E r_o r_\pi + g_m r_{ex} r_o r_\pi}\right), \quad (7.7)$$

which if approximated is equal to eq. (7.2), that is

$$A_f = -\frac{R_C}{R_E}\left(1 - \frac{1}{1 + g_m R_E}\right). \quad (7.8)$$

# Chapter 8

Symbolic Distortion Analysis[1]

This chapter presents a new approach based on the use of describing functions to obtain approximate symbolic expressions for the fundamental frequency component as well as for the second- and third-order harmonics. A single soft or hard transconductance nonlinearity embedded in an arbitrary linear network is studied. Three examples are considered.

The analysis and design of integrated circuits depend on the application of suitable models for the circuit elements, e.g. transistor models. An ideal transistor model is appropriate for initial synthesis, while a nonlinear transistor model may be required to analyse the actual realization of the circuit. The small-signal schematic of a circuit cannot be used directly to estimate the amount of distortion generated by an amplifier, since this schematic is inherently linear. The large-signal behaviour, on the other hand, is very interesting for circuit elements that are nonlinear by concept (e.g. mixers) and it does also describe the signal degradation found in circuits that are supposedly linear, such as audio amplifiers [3], [16], [17].

Total harmonic distortion (THD) for an amplifier is specified as a numerical value. This value does not give any indication of neither the nature of the distortion nor the cause of it. However, by obtaining symbolic expressions for the harmonics and the intermodulation products, a better understanding of

---

1. Parts of this chapter have been published in H. Floberg and S. Mattisson, "Symbolic Distortion Analysis of Nonlinear Elements in Feedback Amplifiers Using Describing Functions", *Int. J. Circuit Theory and Applications*, vol. ICTACV-23, no. 4, pp. 345-356, Jul.-Aug. 1995. Copyright ©1995 John Wiley & Sons Limited. Reproduced with permission.

the distortion mechanism can be obtained. The Volterra series technique [42], [43], [57] can be used to analyse soft nonlinearities for small signal levels but fails when the nonlinearities become hard [32]. The describing function method [5], [18], [62] is an extended version of the frequency response method used for linear networks which is appropriate for soft as well as hard nonlinearities. This method is analytical and similar to the numerical harmonic balance method [33], [41]. The describing function method can be extended to handle multiple nonlinearities, but we will assume that we have a single nonlinearity or that we can lump the nonlinearities together as a single nonlinearity.

Describing functions are often used to study nonlinear behaviour in control theory. However, the method is usually applied to negative feedback amplifiers where the feedback network does not interact with (i.e. load) the amplifier, and normally only the fundamental frequency component is studied. We have adapted the describing function method to the requirements imposed by circuit theory, i.e. we use a more general feedback model which allows for pre-gain, post-gain and a direct path [59], [60], [15]. Second- and third-order harmonic distortion components are considered and the method is used to obtain approximate symbolic expressions for the harmonics, with harmonic balance used to validate the approximations. The results are compared with those obtained with numerical large-signal analysis to verify that the influence of harmonics of higher orders is insignificant. The computer programs CASCA [26] (linear analysis), Mathematica [70] and Maple [12] (nonlinear analysis) were used to do the symbolic and numerical calculations that were necessary to verify different aspects of the method.

8.1 Nonlinear Active Networks

Earlier we have considered linear networks and nonlinear networks that can be linearized at an operating point. The concept of linearization presumes that all nonlinearities are soft and that all signals are small. One important property of a linear network is that it does not generate any distortion. A nonlinear network, on the other hand, generates an amount of distortion which in some way depends on the amplitude of the input signal. In this section we will discuss the influence of negative feedback on distortion and describe some numerical methods to calculate this effect. Further, in accordance with our earlier discussions regarding linear networks, we will assume that the nonlinearity is a voltage-controlled current source.

8.1.1 Basic Feedback Model

In this section we will apply basic feedback theory to a single-transistor circuit. Consider the open-loop transconductance amplifier in figure 8.1.

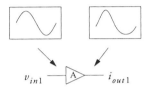

Figure 8.1: Amplifier.

We assume that the amplifier is a bipolar transistor. Thus the output current is

$$i(u) = I_S e^{u/V_T} = I_S e^{(u_0 + u_S)/V_T} = I_C e^{u_S/V_T}, \quad (8.1)$$

where u_0 is the bias level, u_S is the signal component at the input and I_C is the quiescent current at the output. The gain becomes

$$A = \frac{di}{du} = \frac{I_C}{V_T} e^{u_S/V_T}. \quad (8.2)$$

The effects of the nonlinear gain is clearly visible in figure 8.1. For a sinusoidal input signal with $\hat{u}_S = 0.8 V_T$ and $u_0 = 0.67$ the output collector current contains 19 percent second-harmonic distortion and 2.6 percent third-harmonic distortion. As we can see the output signal is rather distorted even for small input levels.

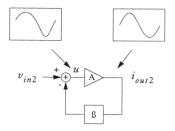

Figure 8.2: Feedback amplifier.

A common way to reduce distortion is to introduce negative feedback as in figure 8.2. The gain is given by the familiar formula for feedback gain,

$$A_f = A/(1 + \beta A). \tag{8.3}$$

Thus to make the distortion figures comparable (i.e. the output levels equal) we have to increase the input signal v_{in2} by a factor equal to the return difference

$$F = 1 + \beta A. \tag{8.4}$$

This means that the fundamental frequency component of the error signal u will have the same level as the previous input signal v_{in1}. The feedback network β is a transresistance. For $F = 10$ the output distortion is reduced to 2.0 percent second-harmonic distortion and 0.43 percent third-harmonic distortion. The error signal, on the other hand, now contains an "inverse" distortion with 18 percent second-harmonic distortion and 3.9 percent third-harmonic distortion. Thus the harmonics introduced in the error signal by the feedback network reduces the effects of the amplifier's nonlinear behaviour, resulting in a less distorted output signal. The price we pay for lower distortion is a loss of gain.

8.1.2 Large-Signal Analysis

The graphs in figure 8.2 were obtained with a numerical large-signal analysis method, where the feedback amplifier was fed with the input reference signal

$$r(t) = r_0 + r_1 \sin \omega t, \tag{8.5}$$

8.1. Nonlinear Active Networks

which was sampled during one period at a number of equidistant time points, see figure 8.3.

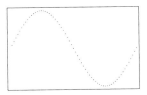

Figure 8.3: Sampled input signal $r(t)$.

We may solve the equation

$$g(u) = u + \beta i_{out}(u) - r = 0 \qquad (8.6)$$

for each time point to obtain the error signal $u(t)$ in figure 8.4.

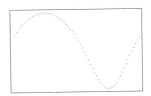

Figure 8.4: Sampled error signal $u(t)$.

The output signal is easily calculated when the error signal is known. Fourier transformation is applied to the waveforms to obtain numerical figures for the harmonic distortion.

The large-signal analysis method can be generalized to handle multiple nonlinear transconductances. Then we need to solve a set of equations

$$YV = I. \qquad (8.7)$$

All linear and linearized circuit elements are included in the admittance matrix Y while the nonlinear current sources are added to the current vector I.

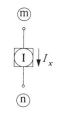

Figure 8.5: Nonlinear current source.

The current contribution from the nonlinear current source in figure 8.5 is

$$\begin{cases} I_m = -(I_x - I_0) \\ I_n = I_x - I_0 \end{cases}, \tag{8.8}$$

where I_x is the total current and I_0 is the bias current. This method makes it possible to study different combinations of linear and nonlinear current sources in order to locate the origin of the distortion.

8.1.3 Harmonic Balance

Harmonic balance is a numerical method which operates in the frequency domain [33], [41]. As the name implies the idea is to balance the harmonics. A sinusoidal input signal to the feedback amplifier in figure 8.2 will generate some amount of harmonics at the output. The harmonics will be fed back to the input where they will generate new harmonics and cross products. All these harmonic components must find their balance. The approximation we have to make is that the number of harmonics considered in the error signal must be limited. Assume that we can neglect the influence of harmonics of the order four and higher. This approximation is referred to as third-order harmonic balance. The error signal is then given by

$$u(t) = u_0 + u_S(t), \tag{8.9}$$

where

$$u_S(t) = u_1 \sin(\omega t + q_1) + u_2 \sin(2\omega t + q_2) + u_3 \sin(3\omega t + q_3), \tag{8.10}$$

8.1. Nonlinear Active Networks

or, in complex notation,

$$u_S = Im(z_1 e^{i\omega t}) + Im(z_2 e^{i2\omega t}) + Im(z_3 e^{i3\omega t}), \qquad (8.11)$$

where

$$z_k = u_k e^{iq_k}. \qquad (8.12)$$

Apply the input signal $r(t)$ found in eq. (8.5). Then we need to solve the following set of harmonic equations

$$\begin{cases} u_0 + \beta(0) N_0^* - r_0 = 0 \\ z_1 + \beta(i\omega) N_1^* - r_1 = 0 \\ z_2 + \beta(i2\omega) N_2^* = 0 \\ z_3 + \beta(i3\omega) N_3^* = 0 \end{cases}, \qquad (8.13)$$

where

$$N_k^* = y_k e^{i\varphi_k} \qquad (8.14)$$

is the k th order harmonic at the output in complex representation. Since u_0 is the bias component, e.g. the base-emitter voltage for a transistor, we will assume that u_0 is known and only solve the three remaining harmonic equations. Thus our problem is to solve

$$\begin{cases} f_1(z_1, z_2, z_3) = 0 \\ f_2(z_1, z_2, z_3) = 0, \\ f_3(z_1, z_2, z_3) = 0 \end{cases} \qquad (8.15)$$

or, written in compact form,

$$f(z) = 0. \qquad (8.16)$$

This system of nonlinear equations can be solved iteratively with the Newton-Raphson algorithm

$$z^{n+1} = z^n - M^{-1} f(z^n), \tag{8.17}$$

where

$$M = \begin{bmatrix} \dfrac{\partial f_1}{\partial z_1} & \dfrac{\partial f_1}{\partial z_2} & \dfrac{\partial f_1}{\partial z_3} \\ \dfrac{\partial f_2}{\partial z_1} & \dfrac{\partial f_2}{\partial z_2} & \dfrac{\partial f_2}{\partial z_3} \\ \dfrac{\partial f_3}{\partial z_1} & \dfrac{\partial f_3}{\partial z_2} & \dfrac{\partial f_3}{\partial z_3} \end{bmatrix}_{|z^n} \tag{8.18}$$

is the Jacobian matrix of the function f. Solving eq. (8.16) for a bipolar transistor with $r_1 = 8 V_T$ and $F = 10$ gives $HD_2 = 18\%$ and $HD_3 = 3.6\%$ for $u(t)$. The slight difference in third-harmonic distortion compared to the figures given at the end of section 8.1.1 is due to the fact that we have neglected higher order harmonics (e.g. $sin^5 \omega t$ affects HD_3).

8.2 Describing Functions

The describing function method [5], [18], [62] is an extended version of the frequency response method used for linear networks. It is an analytical method which is similar to the numerical harmonic balance method. A major difference is that we need analytical expressions (i.e. describing functions) for the harmonics. This method is applicable for soft nonlinearities but can also be used for hard nonlinearities if the feedback network has a lowpass characteristic. Thus the describing function method is appropriate when only a small amount of high-order harmonics is present in the error signal. Distortion analysis based on Volterra series [42], [43], [57] requires that the nonlinear function be approximated with a polynomial function. As we will see in section 8.3.3, low-order polynomial approximations for hard nonlinearities are not sufficient to give acceptable results. Symbolic distortion analyses based on Volterra series assume soft nonlinearities and small signal levels and cannot handle hard nonlinearities [32]. In the context of the Volterra series method, "hard" can be defined as a function that shows "large" deviations from a low-order polynomial within the signal level region of interest.

The different steps involved in the describing function method are described below. The results for some familiar nonlinearities are calculated and compared with those obtained with harmonic balance and large-signal analysis in the following subsections. We will use the variable notation found in appendix C.1.

8.2.1 General Feedback Model

Even though our main objective is to analyse a nonlinear network, we also need a method that allows us to separate the linear parts of the network. Classical control theory requires that the network can be divided into two parts, A and β. In practice this is not possible. Instead we will use the more general feedback model [59], [15] shown in figure 8.6. One important feature of this feedback model is that the familiar basic feedback model can be separated from the linear subnetworks. There are several symbolic analysis methods available that can be used to evaluate the linear parts of the network [32], [25], [49].

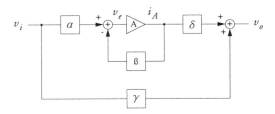

Figure 8.6: General feedback model.

This model is defined by the equations

$$\begin{cases} i_A = Av_e \\ v_e = av_i - \beta i_A = av_i - \beta Av_e \\ v_o = \gamma v_i + \delta i_A = \gamma v_i + \delta Av_e \end{cases} \qquad (8.19)$$

where the parameters are obtained as

$$\alpha = \left.\frac{v_e}{v_i}\right|_{i_A = 0} = \left.\frac{v_e}{v_i}\right|_{A = 0}$$

$$\delta = \left.\frac{v_o}{i_A}\right|_{\substack{v_i = 0 \\ A = 0}}$$

$$\gamma = \left.\frac{v_o}{v_i}\right|_{i_A = 0} = \left.\frac{v_o}{v_i}\right|_{A = 0} \qquad (8.20)$$

$$\beta = -\left.\frac{v_e}{i_A}\right|_{\substack{v_i = 0 \\ A = 0}}$$

8.2.2 Nonlinear Function

We will assume that the nonlinearity is a voltage-controlled current source. The behaviour of the nonlinearity is defined by a nonlinear function $y(u)$. This function must be of such a nature that it is possible to obtain analytical expressions for the Fourier integrals. The nonlinearity may be, for example, piecewise linear or polynomial. Piecewise linear approximations of nonlinearities are widely used in control engineering. Some functions may require a Taylor series expansion, e.g. $\exp(u)$ or $\tanh(u)$. This problem does not occur with the harmonic balance method but is also present with the Volterra series method.

8.2.3 Fourier Transformation

Linear systems are often studied in the frequency domain using the Laplace or Fourier transformation. These methods exploit the periodicity in time. However, the same technique is applicable to nonlinear systems.

8.2. Describing Functions

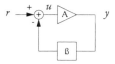

Figure 8.7: Basic feedback amplifier.

Let the nonlinear system in figure 8.7 be forced by a biased sinusoidal reference signal

$$r(t) = r_0 + r_1 \sin \omega t. \tag{8.21}$$

Assume that the input signal to the nonlinear element can be represented by a series of the form

$$u(t) = u_0 + \sum_{k=1}^{n} u_k \sin(k\omega t + q_k). \tag{8.22}$$

It is essential that the same harmonics are present in eqs. (8.22) and (8.23). Thus we neglect harmonics at frequencies higher than nf and approximate the output signal as

$$y(t) = y_0 + \sum_{k=1}^{n} y_k \sin(k\omega t + \varphi_k). \tag{8.23}$$

This function can be rewritten as

$$y(t) = c_0 + \sum_{k=1}^{n} (a_k \cos k\omega t + b_k \sin k\omega t), \tag{8.24}$$

where the coefficients are given by

$$c_0 = \frac{1}{2\pi} \int_0^{2\pi} y(t)\,d(\omega t)$$

$$a_k = \frac{1}{\pi} \int_0^{2\pi} y(t)\cos k\omega t\,d(\omega t). \qquad (8.25)$$

$$b_k = \frac{1}{\pi} \int_0^{2\pi} y(t)\sin k\omega t\,d(\omega t)$$

8.2.4 Harmonic Equations

The describing functions are defined as

$$\begin{cases} N_0 = \dfrac{y_0}{u_0} = \dfrac{c_0}{u_0} \\[1ex] \ldots \\[1ex] N_k = \dfrac{y_k e^{i\varphi_k}}{u_k e^{iq_k}} = \dfrac{b_k + ia_k}{u_k e^{iq_k}}, \\[1ex] \ldots \\[1ex] N_n = \dfrac{y_n e^{i\varphi_n}}{u_n e^{iq_n}} = \dfrac{b_n + ia_n}{u_n e^{iq_n}} \end{cases} \qquad (8.26)$$

where c_0, a_k and b_k are given by eqs. (8.25). In the linear case we would describe the behaviour of the amplifier in figure 8.7 as

$$(1 + \beta A)u - r = 0. \qquad (8.27)$$

8.2. Describing Functions

For a nonlinear amplifier we need one equation similar to eq. (8.27) for each harmonic. Thus we have the following set of harmonic equations

$$\begin{cases} [1 + \beta(0)N_0]u_0 - r_0 = 0 \\ [1 + \beta(i\omega)N_1]u_1 e^{iq_1} - r_1 = 0 \\ \quad \cdots \\ [1 + \beta(in\omega)N_n]u_n e^{iq_n} = 0 \end{cases} \qquad (8.28)$$

It should be noted that each describing function is a function of the bias level and all the harmonics present in the error signal, i.e.

$$N_k = g_k(u_0, u_1, \ldots, u_n, q_1, \ldots, q_n). \qquad (8.29)$$

The output harmonics are given by

$$y_k \sin(k\omega t + \varphi_k) = Im[N_k u_k e^{i(k\omega t + q_k)}]. \qquad (8.30)$$

8.2.5 Harmonic Approximation

Since u_0 is the bias component, e.g. the base-emitter voltage of a transistor, we will assume that u_0 is known and only solve the three remaining harmonic equations. Thus our problem is to solve

$$\begin{cases} f_1(z_1, z_2, z_3) = 0 \\ f_2(z_1, z_2, z_3) = 0, \\ f_3(z_1, z_2, z_3) = 0 \end{cases} \qquad (8.31)$$

or, written in compact form,

$$f(z) = 0, \qquad (8.32)$$

where

$$z_k = u_k e^{iq_k}. \qquad (8.33)$$

The harmonic balance method can be used to obtain numerical values for all z_k in eqs. (8.31). Each harmonic equation can be written as a sum of terms

$$f_k(z) = \sum_{i=1}^{m_k} a_i(z) = 0, \qquad (8.34)$$

where we can use the numerical values for z_k to find the dominant terms. All terms, when multiplied by a user-specified factor, smaller than the largest term are neglected. The set of equations becomes easier to solve when insignificant terms have been removed. We may have a frequency-dependent feedback network. The complex frequency variable s can then be given a fixed numerical value or a range of values to be used for the approximations. The harmonic approximation method is best illustrated by the detailed example of the parabolic function in section 8.3.1.

8.2.6 Summary

1) Use the general feedback model to isolate the basic feedback model.
2) Define a nonlinear function $y(u)$ that describes the amplifier A.
3) Taylor expand $y(u)$ if necessary.
4) Define the reference signal $r(t)$ and the error signal $u(t)$.
5) Calculate the Fourier coefficients c_0, a_k and b_k for $y(u(t))$.
6) Calculate the describing functions N_k from the Fourier coefficients.
7) Set up the harmonic equations.
8) Solve the error signal u numerically with harmonic balance.
9) Use the numerical values for the circuit parameters, the complex frequency s and the error signal u to obtain approximate symbolic expressions for u.
10) Calculate approximate expressions for the output signal y.
11) Calculate the harmonic distortion HD_2 and HD_3.

8.3 Examples

We have considered two relatively soft nonlinearities, the nonlinear parabolic and exponential transconductance functions [6], [7] applicable to the MOS and the bipolar transistor respectively, as well as the hard saturation nonlinearity. Hand calculations have been performed earlier for a transistor in a

8.3. Examples

dedicated test circuit, while the general feedback model allows for the nonlinearity to be embedded in an arbitrary linear network.

8.3.1 Parabolic Function

The nonlinear behaviour for the MOS transistor is defined by

$$y = K(u - V_T)^2. \tag{8.35}$$

We will consider the second- and third-order harmonic distortion in this example. Thus the error signal is given by

$$u(t) = u_0 + u_1 \sin(\omega t + q_1) + u_2 \sin(2\omega t + q_2) + \\ + u_3 \sin(3\omega t + q_3) \tag{8.36}$$

and the output signal becomes

$$y(t) = K[u_0 + u_1 \sin(\omega t + q_1) + u_2 \sin(2\omega t + q_2) + \\ + u_3 \sin(3\omega t + q_3) - V_T]^2 \tag{8.37}$$

The Fourier coefficients for the fundamental frequency are then given by

$$\begin{cases} a_1 = \dfrac{1}{\pi} \displaystyle\int_0^{2\pi} K[u_0 + u_1 \sin(\omega t + q_1) + u_2 \sin(2\omega t + q_2) + \\ \qquad + u_3 \sin(3\omega t + q_3) - V_T]^2 \cos\omega t\, d(\omega t) = \\ \qquad = 2K(u_0 - V_T)u_1 \sin q_1 + K u_1 u_2 \cos(q_1 - q_2) + \\ \qquad + K u_2 u_3 \cos(q_2 - q_3) \\ b_1 = \dfrac{1}{\pi} \displaystyle\int_0^{2\pi} K[u_0 + u_1 \sin(\omega t + q_1) + u_2 \sin(2\omega t + q_2) + \\ \qquad + u_3 \sin(3\omega t + q_3) - V_T]^2 \sin\omega t\, d(\omega t) = \\ \qquad = 2K(u_0 - V_T)u_1 \cos q_1 + K u_1 u_2 \sin(q_1 - q_2) + \\ \qquad + K u_2 u_3 \sin(q_2 - q_3) \end{cases} \tag{8.38}$$

The describing function for the fundamental frequency is

$$N_1 = (2K(u_0 - V_T)u_1 \cos q_1 + K u_1 u_2 \sin(q_1 - q_2) +$$
$$+ K u_2 u_3 \sin(q_2 - q_3) + i[2K(u_0 - V_T)u_1 \sin q_1 +$$
$$+ K u_1 u_2 \cos(q_1 - q_2) + K u_2 u_3 \cos(q_2 - q_3)])/$$
$$(u_1 e^{iq_1})$$

(8.39)

Similar expressions can be calculated for the second- and third-order harmonics. The describing functions for eq. (8.35) are then given by

$$\begin{cases} N_1 = \dfrac{2K(u_0 - V_T)u_1 e^{iq_1} + iK u_1 u_2 e^{-i(q_1 - q_2)} + iK u_2 u_3 e^{-i(q_2 - q_3)}}{u_1 e^{iq_1}} \\[2mm] N_2 = \dfrac{2K(u_0 - V_T)u_2 e^{iq_2} - \dfrac{1}{2}iK u_1^2 e^{2iq_1} + iK u_1 u_3 e^{-i(q_1 - q_3)}}{u_2 e^{iq_2}} \\[2mm] N_3 = \dfrac{2K(u_0 - V_T)u_3 e^{iq_3} - iK u_1 u_2 e^{i(q_1 + q_2)}}{u_3 e^{iq_3}} \end{cases}$$

.(8.40)

We can identify the term $2K(u_0 - V_T)$, which corresponds to the transconductance of the MOS transistor, in each describing function in eqs. (8.40). It is also possible to see how intermodulation between the harmonics affects the describing functions. Assume without loss of generality that $\beta(ik\omega) = \beta$. The harmonic equation for the fundamental frequency is

$$u_1 e^{iq_1} + 2u_1 \beta K(u_0 - V_T)e^{iq_1} + i u_1 u_2 \beta K e^{-i(q_1 - q_2)} +$$
$$+ i u_2 u_3 \beta K e^{-i(q_2 - q_3)} = r_1$$

(8.41)

and similar equations can be found for the harmonics. We insert the numerical values given by the harmonic balance method (see appendix C.2) for q_k and get

8.3. Examples

$$\begin{cases} u_1 + 2u_1\beta K(u_0 - V_T) - u_1 u_2 \beta K - u_2 u_3 \beta K = r_1 \\ u_2 + 2u_2\beta K(u_0 - V_T) - \frac{1}{2}u_1^2 \beta K - u_1 u_3 \beta K = 0 \\ -u_3 - 2u_3\beta K(u_0 - V_T) + u_1 u_2 \beta K = 0 \end{cases} \quad (8.42)$$

The numerical values for u_k can be used to find the dominant terms in eqs. (8.42), which gives

$$\begin{cases} u_1 + 2u_1\beta K(u_0 - V_T) = r_1 \\ u_2 + 2u_2\beta K(u_0 - V_T) - \frac{1}{2}u_1^2 \beta K = 0 \\ -u_3 - 2u_3\beta K(u_0 - V_T) + u_1 u_2 \beta K = 0 \end{cases} \quad (8.43)$$

Eqs. (8.43) have the exact symbolic solutions

$$\begin{cases} u_1 = \dfrac{r_1}{1 + 2\beta K(u_0 - V_T)} \\ u_2 = \dfrac{\beta K r_1^2}{2(1 + 2\beta K(u_0 - V_T))^3}, \\ u_3 = \dfrac{\beta^2 K^2 r_1^3}{2(1 + 2\beta K(u_0 - V_T))^5} \end{cases} \quad (8.44)$$

which are approximate solutions to eqs. (8.42). We find the factor $(1 + 2\beta K(u_0 - V_T))^{2k-1} = F^{2k-1}$, where F is the return difference with respect to the transconductance, in the denominator for each harmonic in eqs. (8.44). Eqs. (8.30), (8.40) and (8.44) give the output harmonics as

$$\begin{cases} y_1 = \dfrac{2K(u_0 - V_T)r_1}{1 + 2\beta K(u_0 - V_T)} \\[2mm] y_2 = \dfrac{Kr_1^2}{2(1 + 2\beta K(u_0 - V_T))^2} - \dfrac{K^2(u_0 - V_T)\beta r_1^2}{(1 + 2\beta K(u_0 - V_T))^3} \\[2mm] y_3 = \dfrac{K^2 \beta r_1^3}{2(1 + 2\beta K(u_0 - V_T))^4} - \dfrac{K^3(u_0 - V_T)\beta^2 r_1^3}{(1 + 2\beta K(u_0 - V_T))^5} \end{cases} \quad (8.45)$$

Thus the harmonic distortion is given by

$$\begin{cases} HD_2 = \dfrac{y_2}{y_1} = \dfrac{r_1}{4(u_0 - V_T)(1 + 2\beta K(u_0 - V_T))^2} = \dfrac{K y_1}{g_m^2}\dfrac{1}{2F} \\[3mm] HD_3' = \dfrac{y_3}{y_1} = \dfrac{\beta K r_1^2}{4(u_0 - V_T)(1 + 2\beta K(u_0 - V_T))^4} = \left(\dfrac{K y_1}{g_m^2}\right)^2 \dfrac{\beta g_m}{2F^2} \end{cases} \quad (8.46)$$

where

$$\begin{cases} g_m = 2K(u_0 - V_T) \\ F = 1 + \beta g_m \end{cases} . \quad (8.47)$$

It should be noted that third-order harmonic distortion occurs only if feedback is applied to the parabolic function. This is evident from the expression for HD_3 in eqs. (8.46), which contains the feedback factor β in the numerator. It can be shown experimentally that a sufficient amount of feedback is required to reduce the third-order harmonic distortion [6]. The results are given in tables 8.1 and 8.2.

8.3. Examples

Table 8.1: Comparison for the MOS transistor between the numerical values for error signal distortion obtained with the different methods.

Harmonic number	Large-signal analysis	Harmonic balance	Describing functions
2	4.10%	4.10%	4.06%
3	0.337%	0.334%	0.330%

Table 8.2: Comparison for the MOS transistor between the numerical values for output signal distortion obtained with the different methods.

Harmonic number	Large-signal analysis	Harmonic balance	Describing functions
2	1.06%	1.06%	1.04%
3	0.0868%	0.0861%	0.0846%

8.3.2 Exponential Function

The nonlinear behaviour for the bipolar transistor is defined by

$$y = I_S e^{u/V_T}. \tag{8.48}$$

We have to use a Taylor series expansion for the exponential function in order to obtain the Fourier coefficients

$$e^x = 1 + x + \frac{1}{2}x^2 + \frac{1}{6}x^3. \tag{8.49}$$

Thus

$$y = I_C \left(1 + \frac{u_S}{V_T} + \frac{1}{2}\left(\frac{u_S}{V_T}\right)^2 + \frac{1}{6}\left(\frac{u_S}{V_T}\right)^3\right). \tag{8.50}$$

The describing functions for eq. (8.50) are given by

$$\begin{cases} N_1 = \left(V_T^2 u_1 e^{iq_1} + \frac{1}{8} u_1^3 e^{iq_1} + \frac{1}{2} i V_T u_1 u_2 e^{-i(q_1-q_2)} + \right. \\ \qquad + \frac{1}{4} u_1 u_2^2 e^{iq_1} + \frac{1}{4} u_1 u_3^2 e^{iq_1} - \frac{1}{8} u_1^2 u_3 e^{-i(2q_1-q_3)} + \\ \qquad \left. + \frac{1}{2} i V_T u_2 u_3 e^{-i(q_2-q_3)} + \frac{1}{8} u_2^2 u_3 e^{i(2q_2-q_3)} \right) \frac{I_C}{V_T^3} \frac{1}{u_1 e^{iq_1}} \\ N_2 = \left(-\frac{1}{4} i V_T u_1^2 e^{i2q_1} + \frac{1}{4} u_1^2 u_2 e^{iq_2} + \frac{1}{2} i V_T u_1 u_3 e^{-i(q_1-q_3)} + \right. \\ \qquad + \frac{1}{4} u_1 u_2 u_3 e^{i(q_1-q_2+q_3)} + V_T^2 u_2 e^{iq_2} + \frac{1}{8} u_2^3 e^{iq_2} + \\ \qquad \left. + \frac{1}{4} u_2 u_3^2 e^{iq_2} \right) \frac{I_C}{V_T^3} \frac{1}{u_2 e^{iq_2}} \\ N_3 = \left(-\frac{1}{24} u_1^3 e^{i3q_1} - \frac{1}{2} i V_T u_1 u_2 e^{i(q_1+q_2)} + \right. \\ \qquad + \frac{1}{8} u_1 u_2^2 e^{-i(q_1-2q_2)} + \frac{1}{4} u_1^2 u_3 e^{iq_3} + \frac{1}{4} u_2^2 u_3 e^{iq_3} + \\ \qquad \left. + V_T^2 u_3 e^{iq_3} + \frac{1}{8} u_3^3 e^{iq_3} \right) \frac{I_C}{V_T^3} \frac{1}{u_3 e^{iq_3}} \end{cases} \quad (8.51)$$

The expressions in eqs. (8.51) are more complicated than the describing functions in eqs. (8.40) for the MOS transistor owing to the Taylor expansion. The transconductance for the bipolar transistor, I_C/V_T, is identified in each expression in eqs. (8.51). If $\beta(ik\omega) = \beta$, then eqs. (8.28) give the approximate symbolic solutions

8.3. Examples

$$\begin{cases} u_1 = \dfrac{r_1}{1+\beta g_m} \\ u_2 = \dfrac{\beta g_m r_1^2}{4(1+\beta g_m)^3 V_T} \\ u_3 = \dfrac{\beta g_m(1-2\beta g_m)r_1^3}{24(1+\beta g_m)^5 V_T^2} \end{cases} \qquad (8.52)$$

where

$$g_m = I_C/V_T. \qquad (8.53)$$

We can see the factor $(1+\beta g_m)^{2k-1} = F^{2k-1}$ in the denominator for each expression in eqs. (8.52). The harmonics at the output are given by

$$\begin{cases} y_1 = \dfrac{g_m r_1}{1+\beta g_m} \\ y_2 = \dfrac{g_m r_1^2}{4(1+\beta g_m)^3 V_T} \\ y_3 = \dfrac{g_m(1-2\beta g_m)r_1^3}{24(1+\beta g_m)^5 V_T^2} \end{cases} \qquad (8.54)$$

If we compare eqs. (8.52) and (8.54), we get

$$\begin{cases} y_1 = g_m u_1 \\ y_2 = \dfrac{u_2}{\beta} \\ y_3 = \dfrac{u_3}{\beta} \end{cases} \qquad (8.55)$$

The harmonic distortion is given by

$$\begin{cases} HD_2 = \dfrac{y_2}{y_1} = \dfrac{r_1}{4(1+\beta g_m)^2 V_T} = \dfrac{y_1}{I_C}\dfrac{1}{4F} \\ HD_3 = \dfrac{y_3}{y_1} = \dfrac{(1-2\beta g_m)r_1^2}{24(1+\beta g_m)^4 V_T^2} = \left(\dfrac{y_1}{I_C}\right)^2 \dfrac{1-2\beta g_m}{24F^2} \end{cases} \quad (8.56)$$

with

$$F = 1 + \beta g_m. \quad (8.57)$$

According to eqs. (8.56), HD_3 can be cancelled by a suitable feedback factor β. In fact, a local minimum for the third-harmonic distortion does exist and can be found experimentally [7]. The results are given in tables 8.3 and 8.4.

Table 8.3: Comparison for the bipolar transistor between the numerical values for error signal distortion obtained with the different methods.

Harmonic number	Large-signal analysis	Harmonic balance	Describing functions
2	7.95%	7.94%	8.00%
3	0.735%	0.724%	0.747%

Table 8.4: Comparison for the bipolar transistor between the numerical values for output signal distortion obtained with the different methods.

Harmonic number	Large-signal analysis	Harmonic balance	Describing functions
2	1.98%	1.98%	2.00%
3	0.183%	0.180%	0.187%

8.3.3 Saturation Function

The nonlinear behaviour of the saturation function is defined by

$$y = \begin{cases} -a, & -1 < u \leq -\dfrac{a}{K} \\ Ku, & -\dfrac{a}{K} < u < \dfrac{a}{K} \\ a, & \dfrac{a}{K} \leq u < 1 \end{cases}. \tag{8.58}$$

This function represents a hard nonlinearity and the third- and fifth-order polynomial mean-square approximations, in figure 8.8, fit poorly. The second and third columns of tables 8.5 and 8.6 show that both these polynomial approximations, as expected, result in insufficient accuracy of the distortion estimations. The Volterra series method relies on such polynomial approximations of the nonlinear function and will thus suffer from large errors.

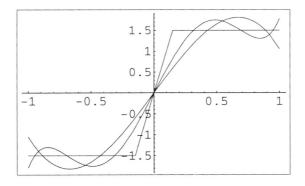

Figure 8.8: Third- and fifth-order polynomial approximations of the saturation function.

The behaviour of the saturated feedback amplifier in figure 8.9 at the output is rather obvious. It has the gain

$$A_f = \frac{K}{1 + \beta K}, \tag{8.59}$$

but saturates for $|y| = a$. However, we will take a closer look at the error signal u.

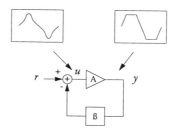

Figure 8.9: Saturated feedback amplifier.

The describing functions for eq. (8.58) are given by

$$\begin{cases} N_1 = \left(4a\cos\sigma + Ku_1 e^{-iq_1}(2e^{i2q_1}\sigma - \sin 2\sigma) + \right. \\ \qquad - \frac{1}{2} iKu_3 e^{i(q_3 - 2\sigma)}(e^{i4\sigma} - 1) + \\ \qquad \left. + \frac{1}{4} iKu_3 e^{-i(q_3 + 4\sigma)}(e^{i8\sigma} - 1)\right) \frac{1}{\pi} \frac{1}{u_1 e^{iq_1}} \\ N_2 = \left(4Ku_2 e^{iq_2}\sigma + 4iKu_0 \sin 2\sigma + \right. \\ \qquad \left. - Ku_2 e^{-iq_2} \sin 4\sigma\right) \frac{1}{2\pi} \frac{1}{u_2 e^{iq_2}} \\ N_3 = \left(16a\cos 3\sigma - 6iKu_1 e^{i(q_1 - 2\sigma)}(e^{i4\sigma} - 1) + \right. \\ \qquad + 3iKu_1 e^{-i(q_1 + 4\sigma)}(e^{i8\sigma} - 1) + 24Ku_3 e^{iq_3}\sigma + \\ \qquad \left. + 2iKu_3 e^{-i(q_3 + 6\sigma)}(e^{i12\sigma} - 1)\right) \frac{1}{12\pi} \frac{1}{u_3 e^{iq_3}} \end{cases} \qquad (8.60)$$

where $\omega t = \sigma$ for $Ku = a$, and $0 \leq \sigma \leq \pi/2$. The nonlinearity is piecewise linear and it is possible to get exact symbolic solutions, i.e. the expressions correspond to the numerical values obtained with the harmonic balance

8.3. Examples

method. However, the trigonometry makes the solutions rather involved and we will therefore replace σ with its numerical value. If $\beta(ik\omega) = \beta$, then eqs. (8.28) give the symbolic solutions

$$\begin{cases} u_1 = 0.833\dfrac{a}{K} + \\ \quad + \dfrac{[0.402(9.42 + 5.12\beta K)(31.3\beta K r_1 - 26.1a\beta - 33.8a\beta^2 K)]}{\beta K(118 + 80.3\beta K + 0.408\beta^2 K^2)} \\ u_2 = 0 \\ u_3 = \dfrac{31.3\beta K r_1 - 26.1a\beta - 33.8a\beta^2 K}{118 + 80.3\beta K + 0.408\beta^2 K^2} \end{cases} \qquad (8.61)$$

Obviously u_3 decreases if β increases (i.e. A_f decreases), if r_1 decreases or if a increases, since all these changes will reduce the saturation. Negative feedback does not have any direct effect on distortion due to saturation, since the gain reduction has to be compensated with an increase in the input signal level[1].

Table 8.5: Comparison for the saturation function between the numerical values for error signal distortion obtained with different approximations.

Harmonic number	Large-signal analysis	Large-signal analysis 3rd order polynomial approx.	Large-signal analysis 5th order polynomial approx.	Harmonic balance and describing functions 3rd order	Harmonic balance 5th order
3	23.3%	1.89%	4.29%	20.1%	23.4%
5	6.95%	-	0.472%	-	7.90%

---

1. This problem has been addressed by several commercial amplifier designers who have introduced soft clipping. The idea is to soften the sharp edges found in the right graph in figure 8.9 in order to get a less harsh distortion. A nonlinear feedback network is used to avoid saturation.

Table 8.6: Comparison for the saturation function between the numerical values for output signal distortion obtained with different approximations.

Harmonic number	Large-signal analysis	Large-signal analysis 3rd order polynomial approx.	Large-signal analysis 5th order polynomial approx.	Harmonic balance and describing functions 3rd order	Harmonic balance 5th order
3	10.7%	1.37%	2.23%	9.26%	10.7%
5	3.20%	-	0.246%	-	3.62%

Chapter 9

Switched-Capacitor Networks

This chapter presents a method that makes it possible to analyse switched-capacitor networks in discrete time using compacted nodal analysis in continuous time. Our objective is to perform time-discrete analysis in the z-domain using any symbolic analysis tool (e.g. CASCA [26]) intended for analysis of time-continuous networks in the s-domain. An equivalent analog circuit is used to describe the switched capacitors in the s-domain with $z = s$ instead of $z = e^{sT}$ which gives a simple translation between the two domains. Two switched-capacitor examples are considered: an inverting amplifier and a biquad.

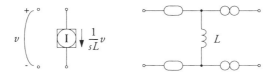

Figure 9.1: (a) Inductive current generator (b) modelled by an inductance and two nullors.

The presented symbolic analysis method is based on a series of papers on nodal analysis of switched-capacitor (SC) networks using the indefinite admittance matrix [44], [45], [40]. We will consider three circuit elements: capacitors, switches and operational amplifiers. Nodal analysis can be used for any network that can be modelled with transadmittances, which may be conductive, capacitive or inductive current generators [25]. The compacted nodal analysis also includes the nullor element [49]. Some programs for symbolic analysis do not directly include inductive current generators, as in figure 9.1 (a). However, the nullor can be used to overcome this problem, see figure 9.1 (b).

9.1 Switched Capacitor

The nodal charge equations are used to describe the behaviour of a time-discrete network. They are based on the charge conservation principle, similar to Kirchhoff's current law, that says: the algebraic sum of charges leaving any node in any phase is zero. The indefinite nodal capacitance matrix C_n can be set up merely by inspection of the network,

$$C_n V_n = Q. \tag{9.1}$$

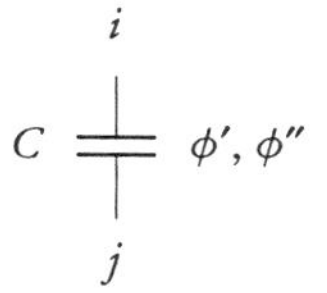

Figure 9.2: Capacitor.

Let us consider a capacitor between the nodes i and j embedded in a switched-capacitor network with n nodes and assume a nonoverlapping clock with two phases ϕ' and ϕ'', see figure 9.2. We need to set up one charge equation for each node in each clock phase,

$$\begin{cases} C(v_i' - v_j') - C(v_i'' - v_j'')z^{-1} = 0 \\ C(v_j' - v_i') - C(v_j'' - v_i'')z^{-1} = 0 \end{cases} \phi' \\ \begin{cases} C(v_i'' - v_j'') - C(v_i' - v_j')z^{-1} = 0 \\ C(v_j'' - v_i'') - C(v_j' - v_i')z^{-1} = 0 \end{cases} \phi'' \tag{9.2}$$

or, written in matrix form,

$$\begin{bmatrix} C & -C & -C/z & C/z \\ -C & C & C/z & -C/z \\ -C/z & C/z & C & -C \\ C/z & -C/z & -C & C \end{bmatrix} \begin{bmatrix} v_i' \\ v_j' \\ v_i'' \\ v_j'' \end{bmatrix} = 0. \tag{9.3}$$

9.2 Equivalent Analog Circuit

The SC-network may be considered as a four-port, see figure 9.3, with four individual transfer functions, one for each combination of clock phases,

$$\begin{bmatrix} v_{out}' \\ v_{out}'' \end{bmatrix} = \begin{bmatrix} H_{11} & H_{12} \\ H_{21} & H_{22} \end{bmatrix} \begin{bmatrix} v_{in}' \\ v_{in}'' \end{bmatrix}, \tag{9.4}$$

where the total output is given by

$$v_{out} = v_{out}' + v_{out}''. \tag{9.5}$$

The equivalent analog nodal admittance equations in the s-domain are

$$\begin{bmatrix} G & -G & -\frac{1}{sL} & \frac{1}{sL} \\ -G & G & \frac{1}{sL} & -\frac{1}{sL} \\ -\frac{1}{sL} & \frac{1}{sL} & G & -G \\ \frac{1}{sL} & -\frac{1}{sL} & -G & G \end{bmatrix} \begin{bmatrix} v_i \\ v_j \\ v_{n+i} \\ v_{n+j} \end{bmatrix} = 0, \tag{9.6}$$

where the variable substitutions are given by

$$\begin{cases} s = z \\ G = C \\ L = 1/C \end{cases}. \qquad (9.7)$$

The schematic for the equivalent analog circuit of the capacitor is shown in figure 9.4. Each capacitor in the z-domain is represented by four elements in the s-domain, where the inductive current generators are used to model the interaction between the two clock phases. A switched-capacitor network with two clock phases and n nodes can be represented by an equivalent analog circuit with $2n - 1$ nodes, if we assume that there is a reference node (i.e. ground) that is common for the two phases.

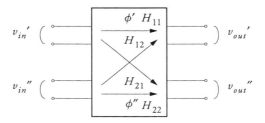

Figure 9.3: Four-port representation of a switched-capacitor network.

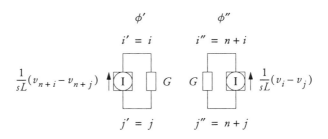

Figure 9.4: Equivalent analog circuit.

9.3 Switches and Operational Amplifiers

A closed switch can be represented by a nullor, as in figure 9.5. A switch that is closed during phase ϕ' or ϕ'' is modelled with a nullor between nodes i and j, or nodes $n+i$ and $n+j$, respectively. The ideal operational amplifier can be modelled by a nullor, see figure 9.6. One nullor is added between the nodes i, j and m, and another nullor is added between the nodes $n+i$, $n+j$ and $n+m$. The operational amplifier with finite gain is conveniently modelled by the block diagram in figure 5.7. Each closed switch or ideal operational amplifier will reduce the rank of the indefinite admittance matrix by one or two, respectively, which gives the total rank

$$N = (2n-1) - (N'_{switch} + N''_{switch}) - 2N_{opamp}. \qquad (9.8)$$

The contraction which is repeatedly applied on the matrix C_n, once for each nullor, in order to obtain the compacted indefinite nodal admittance matrix is explained in section 5.2.

Figure 9.5: (a) Closed switch (b) modelled by a nullor.

Figure 9.6: (a) Ideal operational amplifier (b) modelled by a nullor.

The transfer function from the phase-x input port V_{km} to the phase-y output port V_{ij} is obtained by contracting the rows and columns labelled $k \pm n$ and $m \pm n$ (i.e. short-circuiting the input port for the other phase), and then calculating

$$H_{yx} = \frac{V_{ij}}{V_{km}} = \text{sgn}(k_c - m_c)\text{sgn}(i_c - j_c)(-1)^{k_c + m_c + i_c + j_c} \frac{\left|Y_{i_c j_c}^{k_r m_r}\right|}{\left|Y_{k_c m_c}^{k_r m_r}\right|}, \quad (9.9)$$

where k_r and k_c are the row and column labelled k, respectively (see also section 5.3). Note that the rank of the two matrices for which the determinants are calculated is $N - 3$, where N is given by eq. (9.8).

9.4 Switched-Capacitor Amplifier

The time-continuous equivalent analog circuit method is first illustrated by the analysis of an SC-amplifier example, see figure 9.7. The indefinite nodal admittance matrix (i.e. without nullors) for the equivalent analog network in figure 9.8 is obtained by inspection

$$C_n = \begin{bmatrix} 2G_1 - \frac{2}{sL_1} & 0 & -G_1 + \frac{1}{sL_1} & 0 & 0 & 0 & -G_1 + \frac{1}{sL_1} & 0 & 0 \\ 0 & 0 & 0 & 0 & 0 & 0 & 0 & 0 & 0 \\ -G_1 + \frac{1}{sL_1} & 0 & G_1 & 0 & 0 & 0 & -\frac{1}{sL_1} & 0 & 0 \\ 0 & 0 & 0 & G_2 & -G_2 & 0 & 0 & -\frac{1}{sL_2} & \frac{1}{sL_2} \\ 0 & 0 & 0 & -G_2 & G_2 & 0 & 0 & \frac{1}{sL_2} & -\frac{1}{sL_2} \\ 0 & 0 & 0 & 0 & 0 & 0 & 0 & 0 & 0 \\ -G_1 + \frac{1}{sL_1} & 0 & -\frac{1}{sL_1} & 0 & 0 & 0 & G_1 & 0 & 0 \\ 0 & 0 & 0 & -\frac{1}{sL_2} & \frac{1}{sL_2} & 0 & 0 & G_2 & -G_2 \\ 0 & 0 & 0 & \frac{1}{sL_2} & -\frac{1}{sL_2} & 0 & 0 & -G_2 & G_2 \end{bmatrix}, \quad (9.10)$$

9.4. Switched-Capacitor Amplifier

and the compacted indefinite nodal admittance matrix is, after inserting nullors into eq. (9.10) and contracting,

$$C_n^* = \begin{bmatrix} -G_1 & G_1 & 0 & 0 \\ 0 & 0 & 0 & 0 \\ \frac{1}{sL_1}+G_2 & -\frac{1}{sL_1} & 0 & -G_2 \\ -G_2+G_1-\frac{1}{sL_1} & -G_1+\frac{1}{sL_1} & 0 & G_2 \end{bmatrix}. \qquad (9.11)$$

The row and column contractions due to the nullors have now reduced the rank of the admittance matrix from 9 to 4, see eq. (9.8). The transfer function from the phase-1 input to the phase-2 output is given by

$$H_{21} = -\frac{s^{-1}\frac{1}{L_1}}{G_2} = -z^{-1}\frac{C_1}{C_2}. \qquad (9.12)$$

The relatively large number of nullors present in the equivalent analog network (i.e. compared to an ordinary analog network) makes a larger number of nodes less of a problem. The time-complexity for the calculation of the transfer function for an analog circuit without nullors is $O(n!)$, where n is the number of nodes. However, the use of nullors reduces the rank of the admittance matrix substantially.

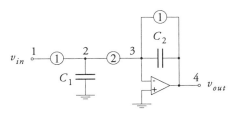

Figure 9.7: Inverting SC-amplifier.

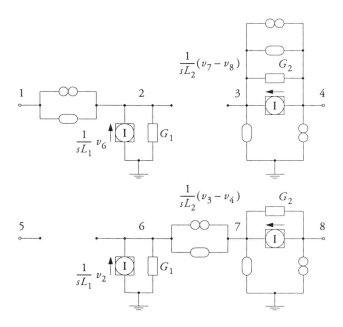

Figure 9.8: Equivalent analog network for the SC-amplifier.

9.5 Switched-Capacitor Biquad

Let us continue with the analysis of the switched-capacitor biquad band-pass filter [24], [30] in figure 9.9. The switches and operational amplifiers reduce the rank of the indefinite admittance matrix from 39 to 7. That is, while the rank of the original matrix is 39, the determinant calculations in eq. (9.9) are performed on matrices of rank 4. The voltage gain is calculated from a phase-1 sampled-and-held input signal to the output sampled at phase 1. Thus, we need to calculate only two of the transfer functions in eq. (9.4), since $v_{out}'' = 0$,

$$\begin{cases} H_{11} = \dfrac{f_1(z)}{g(z)} \\ H_{12} = \dfrac{z f_2(z)}{g(z)} \end{cases}, \qquad (9.13)$$

9.5. Switched-Capacitor Biquad

where

$$\begin{cases} f_1(z) = -C_BC_DC_K + C_AC_EC_K + z^2(-C_AC_BC_G + C_BC_DC_I + \\ \qquad\quad + 2C_BC_DC_K - C_AC_EC_K) - z^4(C_BC_DC_I + C_BC_DC_K) \\ f_2(z) = C_AC_BC_H - C_BC_DC_J - C_AC_EC_K + C_AC_BC_L + \\ \qquad\quad + z^2(C_BC_DC_J + C_AC_EC_K - C_AC_BC_L) \\ g(z) = C_B^2C_D - C_AC_BC_E + z^2(C_AC_BC_C - 2C_B^2C_D + \\ \qquad\quad + C_AC_BC_E - C_BC_DC_F) + z^4(C_B^2C_D + C_BC_DC_F) \end{cases} \quad . (9.14)$$

All the switches change position twice for each input sample [4]. Then,

$$H(z) = \frac{v_{in}(z)\,H_{11}(z^{1/2}) + z^{-1/2}v_{in}(z)\,H_{12}(z^{1/2})}{v_{in}(z)} = \\ = \frac{f_1(z^{1/2}) + f_2(z^{1/2})}{g(z^{1/2})} \quad , \quad (9.15)$$

which gives

$$H(z) = \Big(C_AC_H - C_DC_J - C_DC_K + C_AC_L + \\ \quad + z(-C_AC_G + C_DC_I + C_DC_J + 2C_DC_K - C_AC_L) + \\ \quad - z^2(C_DC_I + C_DC_K)\Big) / \Big(C_BC_D - C_AC_E + \\ \quad + z(C_AC_C - 2C_BC_D + C_AC_E - C_DC_F) + \\ \quad + z^2(C_BC_D + C_DC_F)\Big) \quad . (9.16)$$

A CASCA circuit program for the SC-biquad is found in appendix D.3.

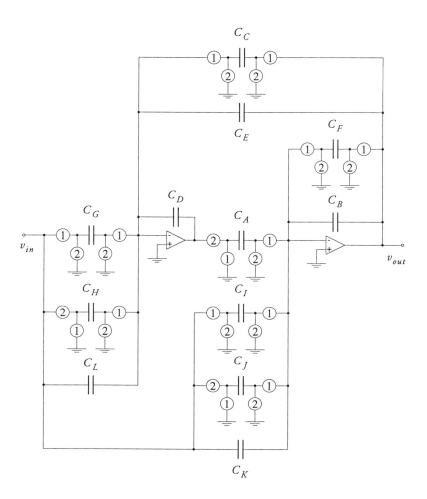

Figure 9.9: SC-biquad of Fleischer and Laker.

Chapter 10

CASCA

This chapter describes a computer aided symbolic circuit analysis program, called CASCA[1], written for the Apple Macintosh personal computer series. The program produces and approximates analytic transfer functions for the small-signal behaviour of linear circuits. It is considered important that the operations performed by the program can be followed by the user. You should not expect to get a simple result without effort for an arbitrary circuit, but if you take the time to write an appropriate description of a circuit schematic and carefully interpret the results from CASCA you have the potential of obtaining a deeper understanding of the behaviour of the circuit.

Analytic transfer functions are extremely useful for the novice as well as for the experienced analog circuit designer. The use of numerical circuit simulation programs in practical analog circuit design is very common nowadays. Undergraduate students are often introduced to such programs, e.g. SPICE, already in fundamental electronics courses. Indeed, numerical simulation may be the only way to handle large networks. The disadvantage, however, with such simulators is that it is difficult to obtain an understanding of a circuit, due to the large number of interacting parameters. Numerical simulators can be used for verifying that a circuit has the intended behaviour, but they do not give any directions on how to improve the performance of the circuit.

The designer can obtain analytic expressions from hand calculations or by using computer aided symbolic circuit analysis methods. Calculations by

---

1. CASCA - Computer Aided Symbolic Circuit Analysis.

hand are only feasible for small circuits, and even then the calculations are tedious and the risk of introducing errors is high. The analytic expressions contain much more useful information than the numerical simulation results do. However, the resulting analytic expressions are generally very long and difficult to interpret, and it is therefore necessary to be able to approximate these expressions guided by the magnitude of the individual parameters.

Circuit design is an iterative process. It is therefore important to have a well-designed user interface, since a significant amount of time will be spent redesigning the circuit. CASCA has a note pad user interface which makes the program well suited for educational purposes. The program contains a text editor, a circuit language interpreter [1], a symbolic circuit evaluator, and an integrated graphics package.

10.1 Language

The circuit program is parsed and executed by the integrated circuit language interpreter. The circuit language used in CASCA is function based, that is all circuit manipulations are performed with function calls. A function result can either be stored in a variable or be used directly as an argument in another function call. The language includes basic calculus, such as the four fundamental rules of arithmetic, trigonometric and exponential functions, and a root solver for polynomial functions. The most important functions are described below.

10.1.1 Symbolic Variables

The symbolic variable types are *mat*, *det* and *detratio*. The variable type *mat* is used to hold a set of equations that describe a small-signal circuit schematic (i.e. the indefinite admittance matrix). The types *det* and *detratio* hold determinants and ratios of determinants, respectively. These three types of symbolic variables are fundamental for CASCA.

10.1.2 Symbolic Functions

The symbolic functions are also fundamental in CASCA. They are used to obtain and approximate symbolic expressions for the small-signal transfer functions, for example the voltage transfer ratio, input impedance and output impedance.

The function *matrix* returns a set of equations for a circuit description. See section 10.2 for further details on how to specify a circuit schematic text.

```
function matrix (x: txt): mat;
```

The function *H* gives the transfer voltage ratio in symbolic form.

```
function H (m: mat; in, out: port): detratio;
```

The function *Z* returns the transfer impedance in symbolic form. The input impedance or the output impedance is obtained if you set the two ports equal.

```
function Z (m: mat; in, out: port): detratio;
```

The function *G* gives the transfer function of a block diagram in symbolic form.

```
function G (m: mat; in, out: int): detratio;
```

The function *det* is used to calculate any minor determinant[1] of the indefinite nodal admittance matrix. The definite nodal admittance matrix is obtained by deleting the row and column corresponding to the reference node[2].

```
function det (m: mat; [row₁, col₁, ..., rowₙ, colₙ:
int]): det;
```

The *simplify* function is used to make the often very long symbolic expressions more understandable. The expressions are shortened based on the magnitude of the individual parameters. This simplification may be valid only in a region near the given values. Thus, it is essential that you specify reasonable

---

1. Note that the cofactor sign is not included.
2. The node number zero is found in row and column one.

values for the circuit elements in the small-signal schematic. Only the low-frequency behaviour is considered if the option *"LF"* is included[1]. Terms that are less than $1/Y$ of the maximum term for each coefficient are ignored if the option *"SHORT=Y"* is applied. Terms that are equal in magnitude may cancel so you have to be careful when you choose Y, but a value between 10 and 100 is often reasonable. The default value is 100. Terms from all coefficients in the polynomial are compared at a fixed frequency (rad/s) when the option *"S=X"* is specified.

```
function simplify (r: det, detratio; ["LF", "S=X",
"SHORT=Y"]): det, detratio;
```

10.1.3 Miscellaneous Functions

The function *port* is used to specify where you intend to apply the input and output signals to your circuit.

```
function port (pos, neg: int): port;
```

10.1.4 Polynomial Functions

The polynomial functions are available to get numerical solutions (i.e. roots) for polynomial equations. Roots can be obtained for a specific symbolic expression guided by the magnitude of the individual circuit elements. These values can be used to decide which approximations that are applicable.

```
function poles (r: detratio): roots;
function zeros (r: detratio): roots;
```

10.2 Circuit Description

The circuit description is specified as a small-signal schematic in a format similar to that used by the well-known numerical circuit simulation program SPICE. The complexity of the symbolic expressions increases rapidly with the number of circuit elements and nodes. The manual transformation from DC to AC schematic makes it possible for the user to simplify the circuit schematic by exploiting the topological knowledge of the circuit. The use of

---

1. Coefficients for s^x where $x \neq 0$ are ignored.

10.2. Circuit Description

dedicated transistor models instead of routine application of general macro models makes it possible to obtain less complex expressions. The number of circuit elements is reduced if circuit elements in parallel, e.g. resistors, are replaced with a single equivalent circuit element.

10.2.1 Circuit Elements

CASCA exploits a few fundamental versatile circuit elements, rather than having a large number of special purpose circuit elements. These fundamental elements are the transconductance, the transcapacitance, the inductive transsusceptance and the nullor, all of which are two-port networks. As a consequence of the choice of fundamental circuit elements, it is also possible to describe directly resistors, capacitors, inductors, short-circuits, ideal transistors, ideal operational amplifiers, ideal operational transconductance amplifiers, nullators and norators.

10.2.2 Transadmittances

Figure 10.1 shows the small-signal schematic for a simple MOS transistor model. A transadmittance becomes an admittance if only two nodes are specified, that is the input port and the output port are assumed to be the same.

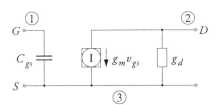

Figure 10.1: Simple MOS transistor model.

```
mos=|
Simple Transistor Model
*
gm   2 3 1 3 50U
gd   2 3 1U
cgs  1 3 50F
*
.END
|
```

The first line in the circuit description, in this example "Simple Transistor Model", is a title card which is ignored. Lines starting with an asterisk are comment cards. The end card should be the last card in the circuit description. The circuit is described with the following transadmittance cards:

```
Gname out+ out- in+ in- value
Rname out+ out- in+ in- value
Cname out+ out- in+ in- value
Lname out+ out- in+ in- value
```

A transadmittance card starts with G, R, C or L followed by a name which is any alphanumeric string. Transadmittances with a name that ends with the number character "#" are interpreted as numerical circuit elements, i.e. the symbolic name is ignored. This is useful if you want to calculate semi-symbolic transfer functions. You have to specify an output port and an input port, or at least one port, for each transadmittance. The card ends with a numerical value that may be a positive integer or floating point number followed by an integer exponent (e.g. 2.65E3) or a scale factor (e.g. 2.65K). The scale factors below are allowed for numbers:

$T=10^{12}$ $G=10^9$ $MEG=10^6$ $K=10^3$ $MIL=25.4 \times 10^{-6}$
$M=10^{-3}$ $U=10^{-6}$ $N=10^{-9}$ $P=10^{-12}$ $F=10^{-15}$

Note that the following two cards are equal, that is R is not a transresistance but a transconductance for which you specify the inverted value.

```
gm  2 3 1 3 50U
rm  2 3 1 3 20K
```

It is important that the numerical values specified for the transadmittances are relevant, since the symbolic expressions for the transfer functions are simplified based on the magnitude of these values.

10.2. Circuit Description

10.2.3 Nullors

A nullor consists of a nullator-norator pair. These elements are called singular network elements. They can be used to analyse network problems, since the ideal transistor and the ideal operational amplifier can be modelled with them. The nullator and norator always occur in pairs.

```
Nname out1 out2 in1 in2
```

A nullor card starts with N followed by a name which is any alphanumeric string. The name may be left out. You have to specify an output port and an input port.

The nullor is useful to model circuits that contain operational amplifiers, for example the active notch filter in figure 10.2. A CASCA example for this filter is found in appendix D.1.

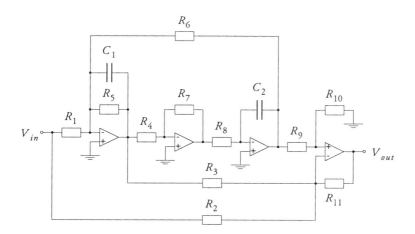

Figure 10.2: Active RC-filter.

The nullor is the equivalent of an ideal operational amplifier. The ideal operational amplifier with single-ended output is obtained by grounding one of the norator nodes. It should be noted that while the operational amplifier has a positive and a negative input, the nullor representation has two equal inputs. The nullor is the equivalent of an ideal transistor if the nullator and the norator have one common node.

10.2.4 Subcircuits

A subcircuit is a group of circuit elements that are inserted wherever the subcircuit is referenced. Subcircuits may contain references to other subcircuits, but nested subcircuit definitions are not allowed.

```
.SUBCKT name localnode₁ … localnodeₙ
…
.ENDS
```

An example of a simple MOS transistor model defined as a subcircuit is found below:

```
.SUBCKT mos 1 2 3
*  G D S
*
gm   2 3 1 3 50U
gd   2 3 1U
cgs  1 3 50F
*
.ENDS
```

All local node numbers must be defined globally when the subcircuit is referenced.

```
Xinstancename globalnode₁ … globalnodeₙ name
```

An example of a reference to the transistor subcircuit is:

```
X1 4 5 6 mos
```

The instance name of the reference will be appended at the end of the names of the inserted subcircuit elements, that is for example:

```
gm_1
gd_1
cgs_1
```

The parameters can be modified when a subcircuit is referenced. An example of this is:

```
X1 4 5 6 mos (gm=100U gd=2U)
```

10.2.5 Block Diagram Elements

CASCA uses two different elements to describe a block diagram, the amplifier and the summation element, see figure 10.3.

Figure 10.3: Block diagram elements.

A circuit is described by a combination of the following two cards:

```
Hname out in (<N(s)>) (<D(s)>)
Sname out in₁ in₂
```

The transfer function of the amplifier card is specified as a numerator and a denominator polynomial. These polynomials may have symbolic or numerical coefficients which are specified in descending order of s, for example:

```
(-1 0) (1 -1)  → N(s)/D(s)=(-s)/(s-1)
(C2=1) (1)     → N(s)/D(s)=C2
```

10.3 Graphics

The *plot, nyquist, nichols* and *rootlocus* commands can be used to get graphical representations of symbolic expressions. We can get graphs of the magnitude and the phase as functions of frequency for the different network functions, for comparison with the output from numerical circuit simulation programs. The axes may be either linear or logarithmic. Note that several graphs may be plotted in the same diagram in order to compare the numerical values of different symbolic expressions.

```
plot (s: float; f₁[,...,f₁₂]: int, float, detratio,
polynomial, roots; x, y: axis; ["TITLE=NAME"]);

nyquist (s: float; f₁[,...,f₁₂]: detratio; w, x, y: axis;
["TITLE=NAME"]);

nichols (s: float; f₁[,...,f₁₂]: detratio; w, x, y: axis;
["TITLE=NAME"]);

rootlocus (g: float; f1[,...,f12]: detratio; k, x, y:
axis; ["TITLE=NAME"]);
```

The function *axis* is used to specify the appearance of graphs generated with the *plot*, *nyquist*, *nichols* and *rootlocus* commands.

```
function axis (min, max: float; ["N=I", "TIC=X", "LOG",
"MAG", "PHASE", "dB", "Hz", "TITLE=NAME"]): axis;
```

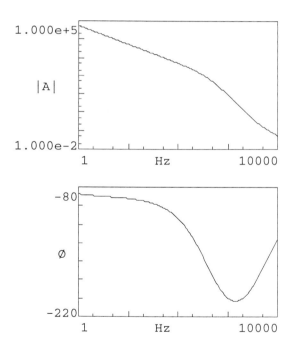

Figure 10.4: (a) Magnitude and (b) phase graphs.

10.3. Graphics

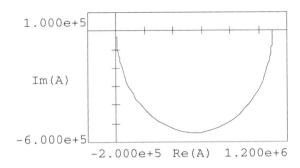

Figure 10.5: Nyquist graph.

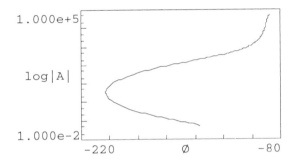

Figure 10.6: Nichols graph.

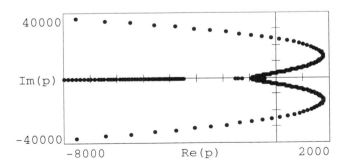

Figure 10.7: Root locus graph.

Chapter 11

Conclusions

The extended pole-splitting method, as presented in chapter 4, can be used to obtain approximate symbolic expressions for poles and zeros. The polynomials of the exact analytic transfer functions are generally extremely long and difficult to interpret. CASCA keeps numerical values for all symbols and uses the magnitude of these values to approximate the exact analytic expressions. CASCA has a built-in numerical root solver, which can be used to find out if the pole-splitting technique is applicable. This technique is extended to handle two closely located roots or a complex conjugated pair of roots. The first-order approximation used by the pole-splitting method is replaced with a second-order approximation, which gives a better estimate of the original polynomial for frequencies close to the two roots. The symbolic roots for the second-order polynomial are approximations of the two roots of the original polynomial.

The nodal analysis (NA) method is a more general method than one often is led to believe when reading textbooks and papers on the subject of network analysis. The usefulness of NA is not limited to $RLC - g_m$ networks. Gyrator transformations can be used for elements without a direct admittance description. The compacted nodal analysis (CNA) method conveniently includes the nullor as a matrix transformation, where each nullor reduces the rank of the matrix by one. The small size of the CNA matrix is an important property since the total number of terms for a symbolic determinant is $O(n!)$. The CNA method also offers the ability to analyse circuit schematics that are mixed with block diagrams.

The feedback amplifiers in section 6.6 can be found in good textbooks on transistor circuit design. We have shown that it is possible to use a systematic method based on the nullor concept to derive all these amplifiers. The simple equivalence rules regarding the nullor make it possible to find different single- and multi-stage transistor configurations that form imperfect implementations of the nullor. Bias aspects, loop gain requirements, linearity considerations and parasitics are some of the characteristics that determine which topology is the most practical to use in a given application.

For some cases it is advantageous to use simpler transistor models for symbolic analysis. The use of simpler models results in less complex symbolic expressions than the use of more refined transistor models, and the user can decide which effects to neglect. This is a major difference between symbolic analysis and numerical analysis, where the results for a refined model become more accurate but not more complicated. The simplest transistor model of them all, the nullor used as a generic transistor, is very useful in circuit design for synthesis as well as analysis. The advantage of the nullor is its ability to model a generic transistor or amplifier without introducing any new symbolic variables. If the approximate symbolic expressions do not correspond to the numerical results of SPICE, we have two options: we may use more complex transistor models or we may want to change the circuit topology.

The describing function method in chapter 8, gives approximate symbolic expressions for the harmonics produced by a nonlinearity which agree well with the numerical results obtained with the harmonic balance method. These symbolic expressions are intended to give a better understanding of the distortion mechanism for a specific nonlinearity. The method relies on the assumption that higher harmonics in the error signal are insignificant. This is ensured if the nonlinearity has a rapidly decaying harmonic spectrum or if the feedback network has a lowpass characteristic. Two soft nonlinearities, the parabolic and the exponential function, and one hard nonlinearity, the saturation function, were successfully analysed with the describing function method. The describing function method and the Volterra series method give similar results when the nonlinear function is approximated with a polynomial (i.e. Taylor series expansion). However, the describing function method gives better results when a piecewise linear approximation is valid.

Chapter 9 presents a method that can be used to perform time-discrete analysis using any symbolic analysis tool intended for analysis of time-continuous networks. An equivalent analog circuit is used to describe the capacitors in the SC-network, where the inductive transsusceptance is used to model the interaction between the two clock phases. Simple variable substitutions allow us to model the time-discrete behaviour using only time-continuous network elements. The nullor is used to model switches and ideal operational amplifiers, where each nullor reduces the rank of the compacted nodal analysis matrix by one.

The symbolic circuit analysis program CASCA, presented in chapter 10, was written to study different aspects of the methods presented in this book. The program gives approximate analytic transfer functions for linear circuits. The purpose of this tool is not merely to calculate these functions automatically, but to give the user the opportunity to interact with the program, and thus create a better understanding of both the circuit behaviour and the calculations involved. CASCA is based on the nodal formulation method, which in general allows only $RLC - g_m$ networks. The equations are obtained if we apply Kirchhoff's current law at each node in a circuit. In CASCA, the nodal analysis is extended to handle the transcapacitance, the inductive transsusceptance and the nullor besides the transconductance.

Appendix A

Determinants

A.1 Definition

We have a matrix A,

$$A = \begin{bmatrix} a_{11} & a_{12} & \cdots & a_{1n} \\ a_{21} & a_{22} & \cdots & a_{2n} \\ \cdot & \cdot & \cdot & \cdot \\ a_{n1} & a_{n2} & \cdots & a_{nn} \end{bmatrix}. \quad (A.1)$$

The determinant $|A|$ can be expressed as a polynomial

$$|A| = \sum (-1)^m a_{1\lambda_1} a_{2\lambda_2} \cdots a_{n\lambda_n}, \quad (A.2)$$

where $\lambda_1, \lambda_2, \ldots, \lambda_n$ is a permutation of the numbers 1, 2, ..., n, and m is the number of inversions in the permutation, i.e. the sign is positive for an even permutation and negative for an odd permutation. The number of inversions is equal to the number of swaps of neighbouring numbers that are necessary to sort a permutation. The determinant is the sum over all permutations. Thus the number of terms in the polynomial is $n!$.

A.2   Minors and Cofactors

As can be seen in eq. (A.2), each term in the determinant polynomial contains one element from each row. Thus the determinant can be written as

$$|A| = a_{i1}\underline{A}_1^i + a_{i2}\underline{A}_2^i + \ldots + a_{in}\underline{A}_n^i, \tag{A.3}$$

where

$$\underline{A}_j^i = (-1)^{i+j}\left|A_j^i\right| \tag{A.4}$$

is called the first-order cofactor and $\left|A_j^i\right|$, called a minor determinant of A, is the determinant for the matrix A with row i and column j deleted. Eq. (A.3) can be rewritten as

$$|A| = \sum_{j=1}^{n} a_{ij}\underline{A}_j^i. \tag{A.5}$$

This expansion of the determinant along a row (or column) is called Laplace's development.

The second-order cofactor is defined as

$$\underline{A}_{ij}^{mn} = (-1)^{m+n+i+j}\left|A_{ij}^{mn}\right|, \tag{A.6}$$

where the minor $\left|A_{ij}^{mn}\right|$ is the determinant for the matrix A with rows m and n, and columns i and j deleted.

A.3   Cramer's Rule

We have a system of n linear equations with n unknowns $x_1, x_2, \ldots, x_n$

$$\begin{cases} a_{11}x_1 + a_{12}x_2 + \ldots + a_{1n}x_n = b_1 \\ a_{21}x_1 + a_{22}x_2 + \ldots + a_{2n}x_n = b_2 \\ \quad\quad\quad\quad\quad\quad\quad\quad\quad\quad\quad\vdots \\ a_{n1}x_1 + a_{n2}x_2 + \ldots + a_{nn}x_n = b_n \end{cases}. \tag{A.7}$$

A.3. Cramer's Rule

Multiplying each equation i with $\underline{A}_k^i$ gives

$$\begin{cases} \underline{A}_k^1 a_{11}x_1 + \underline{A}_k^1 a_{12}x_2 + \ldots + \underline{A}_k^1 a_{1n}x_n = \underline{A}_k^1 b_1 \\ \underline{A}_k^2 a_{21}x_1 + \underline{A}_k^2 a_{22}x_2 + \ldots + \underline{A}_k^2 a_{2n}x_n = \underline{A}_k^2 b_2 \\ \vdots \\ \underline{A}_k^n a_{n1}x_1 + \underline{A}_k^n a_{n2}x_2 + \ldots + \underline{A}_k^n a_{nn}x_n = \underline{A}_k^n b_n \end{cases} \quad (A.8)$$

The sum of these equations is

$$x_1 \sum_{i=1}^{n} a_{i1} \underline{A}_k^i + x_2 \sum_{i=1}^{n} a_{i2} \underline{A}_k^i + \ldots + x_k \sum_{i=1}^{n} a_{ik} \underline{A}_k^i + \ldots \\ \ldots + x_n \sum_{i=1}^{n} a_{in} \underline{A}_k^i = \sum_{i=1}^{n} b_i \underline{A}_k^i \quad (A.9)$$

which can be reduced to

$$x_k \sum_{i=1}^{n} a_{ik} \underline{A}_k^i = x_k |A| = \sum_{i=1}^{n} b_i \underline{A}_k^i = |B|, \quad (A.10)$$

since

$$\sum_{i=1}^{n} a_{ij} \underline{A}_k^i = \begin{cases} |A| & \text{for } j = k \\ 0 & \text{for } j \neq k \end{cases}. \quad (A.11)$$

That is, the determinant for the matrix is equal to zero if column k is replaced with column j. The B matrix is equal to the matrix A with column k replaced with the b_i values. Eq. (A.10) gives Cramer's rule:

$$x_k = \frac{|B|}{|A|} = \frac{\begin{vmatrix} a_{11} & \cdots & a_{1(k-1)} & b_1 & a_{1(k+1)} & \cdots & a_{1n} \\ a_{21} & \cdots & a_{2(k-1)} & b_2 & a_{2(k+1)} & \cdots & a_{2n} \\ \cdot & \cdot & \cdot & \cdot & \cdot & & \cdot \\ a_{n1} & \cdots & a_{n(k-1)} & b_n & a_{2(k+1)} & \cdots & a_{nn} \end{vmatrix}}{|A|} \quad . \tag{A.12}$$

Appendix B

Cubic Polynomial Equation

The fully symbolic solution for a cubic polynomial equation is given in this appendix. The roots for this equation are the poles of the two-stage operational amplifier described in section 4.3.

```
    |\^/|          MAPLE V
._|\|   |/|_.     Copyright (c) 1981-1990 by the University of Waterloo.
 \  MAPLE  /      All rights reserved.  MAPLE is a registered trademark of
 <____  ____>     Waterloo Maple Software.
      |           Type ? for help.
> with(linalg,det);
                                    [det]
>
> solve(
> +4*s^3*(Cc*Cgsl*Cl)/Ri
> +4*s^3*(Cgs6*Cgsl*Cl)/Ri
>
> +2*s^2*(Gml*Cc*Cl)/Ri
> +2*s^2*(Gml*Cgs6*Cl)/Ri
>
> +2*s*(Gm6*Gml*Cc)/Ri
>
> +2*(Gd6*Gdi*Gml)/Ri
> +2*(Gd6*Gdl*Gml)/Ri
> +2*(Gd7*Gdi*Gml)/Ri
> +2*(Gd7*Gdl*Gml)/Ri
> ,s);
                    %3                              %3         1/2
     %6 + %5 - 1/3 ----, - 1/2 %6 - 1/2 %5 - 1/3 ---- + 1/2 3    (%6 - %5) I,
                    %2                              %2

                                %3         1/2
        - 1/2 %6 - 1/2 %5 - 1/3 ---- - 1/2 3    (%6 - %5) I
                                %2
```

Appendix B. Cubic Polynomial Equation

$$\%1 := (-Gm6^2 \, Cc^3 \, Gml^2 \, Cl - Gm6^2 \, Cc^2 \, Gml^2 \, Cgs6 \, Cl + 8 \, Gml^3 \, Gm6^3 \, Cc \, Cgsl$$
$$+ 4 \, Gdi \, Gd6 \, Gml^2 \, Cl^2 \, Cc^2 + 4 \, Gdl \, Gd6 \, Gml^2 \, Cl^2 \, Cc^2 + 4 \, Gdi \, Gd7 \, Gml^2 \, Cl^2 \, Cc^2$$
$$+ 4 \, Gdl \, Gd7 \, Gml^2 \, Cl^2 \, Cc^2 + 4 \, Gdi \, Gd6 \, Gml^2 \, Cl^2 \, Cgs6$$
$$+ 4 \, Gdl \, Gd6 \, Gml^2 \, Cl^2 \, Cgs6 + 4 \, Gdi \, Gd7 \, Gml^2 \, Cl^2 \, Cgs6$$
$$+ 4 \, Gdl \, Gd7 \, Gml^2 \, Cl^2 \, Cgs6 + 8 \, Cc \, Gdi \, Gd6 \, Gml^2 \, Cl \, Cgs6$$
$$+ 8 \, Cc \, Gdl \, Gd6 \, Gml^2 \, Cl \, Cgs6 + 8 \, Cc \, Gdi \, Gd7 \, Gml^2 \, Cl \, Cgs6$$
$$+ 8 \, Cc \, Gdl \, Gd7 \, Gml^2 \, Cl \, Cgs6 - 36 \, Cl \, Gml \, Gm6^2 \, Cc \, Cgsl \, Gdi \, Gd6$$
$$- 36 \, Cl \, Gml \, Gm6 \, Cc^2 \, Cgsl \, Gdl \, Gd6 - 36 \, Cl \, Gml \, Gm6 \, Cc^2 \, Cgsl \, Gdi \, Gd7$$
$$- 36 \, Cl \, Gml \, Gm6 \, Cc^2 \, Cgsl \, Gdl \, Gd7 + 108 \, Cl \, Cc \, Cgsl^2 \, Gdi^2 \, Gd6^2$$
$$+ 108 \, Cl \, Cc \, Cgsl^2 \, Gdl^2 \, Gd6^2 + 108 \, Cl \, Cc \, Cgsl^2 \, Gdi^2 \, Gd7^2$$
$$+ 108 \, Cl \, Cc \, Cgsl^2 \, Gdl^2 \, Gd7^2 + 216 \, Cl \, Cc \, Cgsl^2 \, Gdi \, Gd6 \, Gdl$$
$$+ 216 \, Cl \, Cc \, Cgsl^2 \, Gdi \, Gd6 \, Gd7 + 432 \, Cl \, Cc \, Cgsl^2 \, Gdi \, Gd6 \, Gdl \, Gd7$$
$$+ 216 \, Cl \, Cc \, Cgsl^2 \, Gdl \, Gd6 \, Gd7 + 216 \, Cl \, Cc \, Cgsl^2 \, Gdi \, Gd7 \, Gdl$$
$$+ 108 \, Cl \, Cgs6 \, Cgsl \, Gdi^2 \, Gd6^2 + 108 \, Cl \, Cgs6 \, Cgsl \, Gdl^2 \, Gd6^2$$
$$+ 108 \, Cl \, Cgs6 \, Cgsl \, Gdi^2 \, Gd7^2 + 108 \, Cl \, Cgs6 \, Cgsl \, Gdl^2 \, Gd7^2$$
$$- 36 \, Cl \, Cgs6 \, Gml \, Gm6 \, Cc \, Cgsl \, Gdi \, Gd6 - 36 \, Cl \, Cgs6 \, Gml \, Gm6 \, Cc \, Cgsl \, Gdl \, Gd6$$
$$- 36 \, Cl \, Cgs6 \, Gml \, Gm6 \, Cc \, Cgsl \, Gdi \, Gd7 - 36 \, Cl \, Cgs6 \, Gml \, Gm6 \, Cc \, Cgsl \, Gdl \, Gd7$$
$$+ 216 \, Cl \, Cgs6 \, Cgsl \, Gdi \, Gd6^2 \, Gdl + 216 \, Cl \, Cgs6 \, Cgsl \, Gdi^2 \, Gd6 \, Gd7$$
$$+ 432 \, Cl \, Cgs6 \, Cgsl \, Gdi \, Gd6 \, Gdl \, Gd7 + 216 \, Cl \, Cgs6 \, Cgsl \, Gdl^2 \, Gd6 \, Gd7$$
$$+ 216 \, Cl \, Cgs6 \, Cgsl \, Gdi \, Gd7^2 \, Gdl)^{1/2} \, Gml \, 3^{1/2}$$
$$\Big/ \, (Cgsl^2 \, Cl^{3/2} \, (Cc + Cgs6)^{3/2})$$

$$\%2 := 2\ Cc\ Cgs1\ C1 + 2\ Cgs6\ Cgs1\ C1$$

$$\%3 := Gm1\ Cc\ C1 + Gm1\ Cgs6\ C1$$

$$\%4 := \frac{Gd6\ Gdi\ Gm1 + Gd6\ Gdl\ Gm1 + Gd7\ Gdi\ Gm1 + Gd7\ Gdl\ Gm1}{\%2}$$

$$\%5 := \left(\frac{1}{6} \frac{\%3\ Gm6\ Gm1\ Cc}{\%2^2} - \frac{1}{2} \%4 - \frac{1}{27} \frac{\%3^3}{\%2^3} - \frac{1}{72} \%1 \right)^{1/3}$$

$$\%6 := \left(\frac{1}{6} \frac{\%3\ Gm6\ Gm1\ Cc}{\%2^2} - \frac{1}{2} \%4 - \frac{1}{27} \frac{\%3^3}{\%2^3} + \frac{1}{72} \%1 \right)^{1/3}$$

```
>
> quit;
bytes used=445272, alloc=393144, time=1.016
```

Appendix C

Describing Function Method

C.1 Variable Notation

The following variables are used for the nonlinear describing function method. Their units are given in parentheses.

A	nonlinear transconductance amplifier (Ω^{-1})
α	voltage pre-gain
β	transresistance feedback gain (Ω)
δ	transresistance post-gain (Ω)
F	return difference with respect to A
g_m	linearized transconductance gain (Ω^{-1})
γ	direct path voltage gain
HD_k	kth harmonic distortion at output
N_k	describing function (normalized form) for kth harmonic (Ω^{-1})
r_k	magnitude of kth harmonic of reference signal (V)
$u_k e^{iq_k}$	kth harmonic of error signal in complex form (V)
y	nonlinear current function (A)
$y_k e^{i\varphi_k}$	kth harmonic of output signal in complex form (A)

C.2 Numerical Values

The following numerical solutions were found with the harmonic balance method for the parabolic function example in section 8.3.1:

$$\begin{cases} z_1 = u_1 e^{iq_1} = 0.102384 \angle 0 \\ z_2 = u_2 e^{iq_2} = 0.0041995 \angle 1.5708 \\ z_3 = u_3 e^{iq_3} = 0.00034221 \angle 3.14159 \end{cases} \quad . \quad (C.1)$$

Appendix D

CASCA Examples

This appendix gives a few CASCA circuit program examples [26]. A circuit program starts with a text definition of the circuit, the netlist. This netlist may be considered as a subprogram within the CASCA program. The subprogram has a syntax similar to SPICE [56], and is parsed and converted into an admittance description by the function *matrix*. The circuit is described with transadmittances (i.e. G, L and C) and nullors. The transconductance card, G, may take either two or four nodes, where a card with only two nodes (i.e. a one-port) is a resistor. Each transadmittance card ends with a numerical value which is used when the symbolic results are approximated. The usual scale factors (i.e. prefix) familiar from SPICE are used[1]. The transfer functions H and Z are always given for a two-port with differential ports. These functions require an admittance matrix and two ports as input parameters, and the result is the ratio of two determinants. These symbolic results may be simplified with the function *simplify* or when they are printed. The option *"MATH"* is useful if we want to manipulate the approximated results with general symbolic mathematics programs (e.g. Maple [12] and Mathematica [70]).

---

1. Note that the prefix MEG is used for mega (10^6), and M is used for milli (10^{-3}).

D.1 Active RC-Filter

```
* Computer Aided Symbolic Circuit Analysis.
* Henrik Floberg, Copyright ©1992.
* Active RC filter with nullors as ideal op-amps.
*
RC=|
*
N1 3 0 0 2
N2 5 0 0 4
N3 7 0 0 6
N4 10 0 8 9
*
R1 1 2 1.
R2 1 9 1.
R3 3 9 1.
R4 3 4 1.
R5 2 3 1.
R6 2 7 1.
R7 4 5 1.
R8 5 6 1.
R9 7 8 1.
R10 8 0 1.
R11 9 10 1.
*
C1 2 3 1.
C2 6 7 1.
*
.END
|
*
m=matrix(RC)
*
Vin=port(1,0)
Vout=port(10,0)
*
Hs=H(m,Vin,Vout)
print(Hs,"MATH")
```

D.2 Two-Stage Operational Amplifier

```
* Computer Aided Symbolic Circuit Analysis.
* Henrik Floberg, Copyright ©1992.
* Dept. of Applied Electronics, Lund University, Lund, Sweden.
* Two-stage CMOS op-amp with capacitive load.
*
two=|
*
ri 1 0 1M
ri 2 0 1M
*
*    d s g s
*1
cgdi 1 3 1F
cdbi 3 0 1F
*
gmi 3 0 1 0 20e-6
gdi 3 0 100e-9
*2
cgdi 2 4 1F
cdbi 4 0 1F
*
gmi 4 0 2 0 20e-6
gdi 4 0 100e-9
*3
cgsl 3 0 70F
cdbl 3 0 1F
*
gml 3 0 3 0 15e-6
gdl 3 0 100e-9
*4
cgsl 3 0 70F
cgdl 3 4 1F
cdbl 4 0 1F
*
gml 4 0 3 0 15e-6
gdl 4 0 100e-9
*5
* ideal current generator
*
*6
cgs6 4 0 290F
cdb6 5 0 1F
*
gm6 5 0 4 0 70e-6
gd6 5 0 450e-9
*7
gd7 5 0 410e-9
*
* compensation capacitor
cc 4 5 1p
* load capacitor
cl 5 0 10p
*
.END
|
*
adm=matrix(two)
*
in=port(2,1)
out=port(5,0)
*
```

Appendix D. CASCA Examples

```
Hs=H(adm,in,out)
print(Hs,"INFO","SHORT=100")
*
print(zeros(Hs))
print(poles(Hs))
*
H_approx=simplify(Hs,"SHORT=10")
*
w=axis(1,1e9,"N=200","LOG","TITLE=w (rad/s)")
mag=axis(0,80,"TIC=10","DB","N=100","MAG","LOG","TITLE=A (dB)")
phase=axis(-360,0,"TIC=45","N=100","PHASE","TITLE=Ø (deg)")
*
plot(s,Hs,H_approx,w,mag,"TITLE=Magnitude")
plot(s,Hs,H_approx,w,phase,"TITLE=Phase")
*
print(simplify(Hs,"S=7.0000000e+7","SHORT=10"),"MATH")
print(simplify(Hs,"S=2.1118881e+8","SHORT=10"),"MATH")
*
print(simplify(Hs,"S=2.3470719e+3","SHORT=10"),"MATH")
print(simplify(Hs,"S=5.5321356e+6","SHORT=10"),"MATH")
print(simplify(Hs,"S=1.0564750e+8","SHORT=10"),"MATH")

print(Hs,"INFO","SHORT=100")
1

-> 2*s^4*(Cc*Cgdi^2*Cgsl) (1.4000000e-55)

-> 2*s^3*(Cc*Cgdi*Cgsl)/Ri (1.4000000e-37)

-> -2*s^2*(Gmi*Cc*Cgsl)/Ri (-2.8000000e-27)
2*s^2*(Gml*Cc*Cgdi)/Ri (3.0000000e-29)

2*s*(Gm6*Gmi*Cgsl)/Ri (1.9600000e-19)
-> -2*s*(Gmi*Gml*Cc)/Ri (-6.0000000e-19)

-> 2*(Gm6*Gmi*Gml)/Ri (4.2000000e-11)

time=2.1666667e-1
count=3
subcount=128
elimsubcount=56
writesubcount=7

/

4*s^4*(Cc*Cgdi*Cgs6*Cgsl) (8.1200000e-53)
-> 4*s^4*(Cc*Cgdi*Cgsl*Cl) (2.8000000e-51)
4*s^4*(Cgdi*Cgs6*Cgsl*Cl) (8.1200000e-52)

4*s^3*(Cc*Cgs6*Cgsl)/Ri (8.1200000e-35)
-> 4*s^3*(Cc*Cgsl*Cl)/Ri (2.8000000e-33)
4*s^3*(Cgs6*Cgsl*Cl)/Ri (8.1200000e-34)

4*s^2*(Gm6*Cc*Cgsl)/Ri (1.9600000e-26)
2*s^2*(Gml*Cc*Cgs6)/Ri (8.7000000e-27)
-> 2*s^2*(Gml*Cc*Cl)/Ri (3.0000000e-25)
2*s^2*(Gml*Cgs6*Cl)/Ri (8.7000000e-26)

2*s*(Gdi*Gml*Cl)/Ri (3.0000000e-20)
2*s*(Gdl*Gml*Cl)/Ri (3.0000000e-20)
-> 2*s*(Gm6*Gml*Cc)/Ri (2.1000000e-18)
```

D.2. Two-Stage Operational Amplifier 149

```
4*(Gd6*Gdi*Gdl)/Ri   (1.8000000e-17)
-> 2*(Gd6*Gdi*Gml)/Ri  (1.3500000e-15)
-> 2*(Gd6*Gdl*Gml)/Ri  (1.3500000e-15)
4*(Gd7*Gdi*Gdl)/Ri   (1.6400000e-17)
2*(Gd7*Gdi*Gml)/Ri   (1.2300000e-15)
2*(Gd7*Gdl*Gml)/Ri   (1.2300000e-15)

time=6.1066667e+1
count=14
subcount=11068
elimsubcount=515
writesubcount=19
print(zeros(Hs))
   1.4200000e-55      0
   1.4300000e-37      1
  -2.8398100e-27      2
  -4.0591400e-19      3
   4.2280000e-11      4

         X                   F(X)                 DX
         -                   ----                 --
    7.0000000e+7        -6.3108872e-30       8.1631965e-8      5
   -2.1118881e+8        -3.1554436e-30       3.1240607e-6      3
    2.0000000e+10                    0       3.9720281e+2      2
   -1.0070423e+18                    0       3.9720281e+2      1
END OF ROOTS.
print(poles(Hs))
   3.7969012e-51      0
   3.8117427e-33      1
   4.2379712e-25      2
   2.2287918e-18      3
   5.2288000e-15      4

         X                   F(X)                 DX
         -                   ----                 --
   -2.3470719e+3         3.8518599e-34       2.0879155e-7      2
   -5.5321356e+6        -1.3277361e-30       5.6593869e-3      3
   -1.0564750e+8         1.8294678e-28       1.1117937e-2      1
   -1.0039088e+18                    0       1.1117937e-2      1
END OF ROOTS.
plot(s,Hs,H_approx,w,mag,"TITLE=Magnitude")
plot(s,Hs,H_approx,w,phase,"TITLE=Phase")
```

```
print(simplify(Hs,"S=7.0000000e+7","SHORT=10"),"MATH")
1*
(

-2*s^2*(Gmi*Cc*Cgsl)/Ri

+2*s*(Gm6*Gmi*Cgsl)/Ri
-2*s*(Gmi*Gml*Cc)/Ri

+2*(Gm6*Gmi*Gml)/Ri

)
/
(

+4*s^3*(Cc*Cgsl*Cl)/Ri
+4*s^3*(Cgs6*Cgsl*Cl)/Ri

+2*s^2*(Gml*Cc*Cl)/Ri
+2*s^2*(Gml*Cgs6*Cl)/Ri

+2*s*(Gm6*Gml*Cc)/Ri

)
print(simplify(Hs,"S=2.1118881e+8","SHORT=10"),"MATH")
1*
(

-2*s^2*(Gmi*Cc*Cgsl)/Ri

+2*s*(Gm6*Gmi*Cgsl)/Ri
-2*s*(Gmi*Gml*Cc)/Ri

+2*(Gm6*Gmi*Gml)/Ri

)
/
(

+4*s^3*(Cc*Cgsl*Cl)/Ri
+4*s^3*(Cgs6*Cgsl*Cl)/Ri

+2*s^2*(Gml*Cc*Cl)/Ri
+2*s^2*(Gml*Cgs6*Cl)/Ri

)
print(simplify(Hs,"S=2.3470719e+3","SHORT=10"),"MATH")
1*
(

+2*(Gm6*Gmi*Gml)/Ri

)
/
(

+2*s*(Gm6*Gml*Cc)/Ri

+2*(Gd6*Gdi*Gml)/Ri
+2*(Gd6*Gdl*Gml)/Ri
+2*(Gd7*Gdi*Gml)/Ri
+2*(Gd7*Gdl*Gml)/Ri

)
```

D.2. Two-Stage Operational Amplifier

```
print(simplify(Hs,"S=5.5321356e+6","SHORT=10"),"MATH")
1*
(

+2*(Gm6*Gmi*Gml)/Ri

)
/
(

+2*s^2*(Gml*Cc*Cl)/Ri
+2*s^2*(Gml*Cgs6*Cl)/Ri

+2*s*(Gm6*Gml*Cc)/Ri

)
print(simplify(Hs,"S=1.0564750e+8","SHORT=10"),"MATH")
1*
(

-2*s^2*(Gmi*Cc*Cgsl)/Ri

+2*s*(Gm6*Gmi*Cgsl)/Ri
-2*s*(Gmi*Gml*Cc)/Ri

+2*(Gm6*Gmi*Gml)/Ri

)
/
(

+4*s^3*(Cc*Cgsl*Cl)/Ri
+4*s^3*(Cgs6*Cgsl*Cl)/Ri

+2*s^2*(Gml*Cc*Cl)/Ri
+2*s^2*(Gml*Cgs6*Cl)/Ri

)
```

D.3 SC-Biquad

```
* Computer Aided Symbolic Circuit Analysis.
* Henrik Floberg, Copyright ©1995.
* Switched-capacitor biquad Fleischer & Laker.
*
biquad=matrix(|
* Laker and Fleischer
*
.SUBCKT switch 1 2
*
N 1 2 1 2
*
.ENDS
*
.SUBCKT op 1 3 4 21 23 24 0
*
N 1 0 3 4
N 21 0 23 24
*
.ENDS
*
.SUBCKT cap 2 4 6 8
*
G 2 4 1.
G 6 8 1.
*
L 2 4 8 6 1.
L 6 8 4 2 1.
*
.ENDS
*
.SUBCKT switchcap_1221 1 2 3 4 21 22 23 24 0
*
X 3 4 23 24 cap
X 2 4 switch
X 1 3 switch
X 24 0 switch
X 23 0 switch
*
.ENDS
*
.SUBCKT switchcap_2121 1 2 3 4 21 22 23 24 0
*
X 3 4 23 24 cap
X 4 0 switch
X 1 3 switch
X 22 24 switch
X 23 0 switch
*
.ENDS
*
```

D.3. SC-Biquad

```
X1 9 0 4 29 0 24 0 op
X2 15 0 12 35 0 32 0 op
*
XA 12 9 11 10 32 29 31 30 0 switchcap_2121
XB 12 15 32 35 cap
XC 15 4 8 7 35 24 28 27 0 switchcap_1221
XD 4 9 24 29 cap
XE 4 15 24 35 cap
XF 15 12 14 13 35 32 34 33 0 switchcap_1221
XG 4 1 3 2 24 21 23 22 0 switchcap_1221
XH 4 1 6 5 24 21 26 25 0 switchcap_2121
XI 12 1 17 16 32 21 37 36 0 switchcap_1221
XJ 12 1 19 18 32 21 39 38 0 switchcap_2121
XK 1 12 21 32 cap
XL 1 4 21 24 cap
*
*N 1 0 1 0
N 21 0 21 0
*
.END
|)
*
in=port(1,0)
*
out=port(15,0)
*
H115=H(biquad,in,out)
print(H115,"MATH")
*
```

Bibliography

[1] A. Aho, R. Sethi, J. Ullman, *Compilers principles, techniques, and tools*. Addison-Wesley, 1986.

[2] G. E. Alderson, P. M. Lin, "Computer Generation of Symbolic Network Functions - A New Theory and Implementation", *IEEE Trans. Circuit Theory*, vol. CT-20, no. 1, pp. 48-56, Jan. 1973.

[3] P. E. Allen, "Large-signal performance of audio power amplifiers in the frequency domain", *J. Audio Eng. Society*, vol. 27, pp. 638-646, Sep. 1979.

[4] P. E. Allen, E. Sánchez-Sinencio, *Switched Capacitor Circuits*. Van Nostrand Reinhold, 1984.

[5] D. P. Atherton, *Nonlinear control engineering*. Van Nostrand Reinhold, 1982.

[6] P. J. Baxandall, "Audio power amplifier design: Negative feedback and non-linearity distortion", *Wireless World*, pp. 53-56, Dec. 1978.

[7] P. J. Baxandall, "Audio power amplifier design: More on negative feedback and non-linearity distortion", *Wireless World*, pp. 69-73, Feb. 1979.

[8] H. W. Bode, *Network Analysis and Feedback Amplifier Design*. New York: Van Nostrand, 1945.

[9] H. J. Carlin, D. C. Youla, "Network Synthesis with Negative Resistors", *Proc. IRE*, pp. 907-920, May 1961.

[10] H. J. Carlin, "Singular Network Elements", *IEEE Trans. Circuit Theory*, vol. CT-11, pp. 67-72, March 1964.

[11] A. Carlosena, G. S. Moschytz, "Nullators and norators in voltage to current mode transformations", *Int. J. cir. theor. appl.*, vol. 21, no. 4, pp. 421-424, Jul.-Aug. 1993.

[12] B. W. Char, K. O. Geddes, G. H. Gonnet, M. B. Monagan, S. M. Watt, *MAPLE Reference Manual*. Dept. of Computer Science, University of Waterloo, WATCOM, 1988.

[13] W. K. Chen, *Active Network Analysis*. World Scientific, 1991.

[14] E. M. Cherry, D. E. Hooper, "The design of wide-band transistor feedback amplifiers", *Proc. IEE*, vol. 110, no. 2, pp. 375-389, Feb. 1963.

[15] E. M. Cherry, D. E. Hooper, *Amplifying Devices and Low-Pass Amplifier Design*. New York: Wiley, 1968.

[16] E. M. Cherry, "Transient Intermodulation Distortion — Part I: Hard Nonlinearity", *IEEE Trans. Acoustics, Speech, and Signal Processing*, vol. ASSP-29, no. 2, pp. 137-146, Apr. 1981.

[17] E. M. Cherry, K. P. Dabke, "Transient Intermodulation Distortion — Part 2: Soft Nonlinearity", *J. Audio Eng. Soc.*, vol. 34, no. 1/2, pp. 19-35, Jan./Feb. 1986.

[18] P. A. Cook, *Nonlinear dynamical systems*. Prentice-Hall International, 1986.

[19] M. Degrauwe, "IDAC: An Interactive Design Tool for Analog CMOS Circuits", *IEEE J. Solid-State Circuits*, vol. SC-22, no. 6, pp. 1106-1116, Dec. 1987.

[20] M. Desai, P. Aronhime, "Current-Mode Synthesis Using Node Expansion Techniques", *Analog Integrated Circuits and Signal Processing*, vol. 6, no. 3, pp. 255-263, Nov. 1994.

Bibliography 157

[21] F. V. Fernández, A. Rodríguez-Vázquez, J. L. Huertas, "A Tool for Symbolic Analysis of Analog Integrated Circuits Including Pole/Zero Extraction", *Proc. ECCTD-91*, pp. 752-761, 1991.

[22] F. V. Fernández, A. Rodríguez-Vázquez, J. L. Huertas, "Interactive AC Modeling and Characterization of Analog Circuits via Symbolic Analysis", *Analog Integrated Circuits and Signal Processing*, vol. 1, no. 3, pp. 183-208, Nov. 1991.

[23] J. K. Fidler, J. I. Sewell, "Symbolic Analysis for Computer-Aided Circuit Design - The Interpolative Approach", *IEEE Trans. Circuit Theory*, vol. CT-20, no. 6, pp. 738-741, Nov. 1973.

[24] P. E. Fleischer, K. R. Laker, "A family of active switched capacitor biquad building blocks", *The Bell System Technical Journal*, vol. 58, no. 10, pp. 2235-2269, Dec. 1979.

[25] H. Floberg, *Computer Aided Symbolic Circuit Analysis.* Thesis, Department of Applied Electronics, Lund University, Lund, Sweden, Dec. 1992.

[26] H. Floberg, *'CASCA', Tutorial.* Department of Applied Electronics, Lund University, Lund, Sweden, 1994.

[27] H. Floberg, S. Mattisson, "Computer Aided Symbolic Circuit Analysis", *Alta Frequenza Rivista di Elettronica*, vol. 5, no. 6, pp. 312-316, Nov.-Dec. 1993.

[28] H. Floberg, S. Mattisson, "CASCA", *Symbolic Methods and Applications to Circuit Design, Proc. SMACD'92*, pp. 73-82, Florence, Italy, Oct. 1992.

[29] H. Gaunholt, P. Heikkilä, K. Mannersalo, V. Porra, M. Valtonen, "Gyrator transformation – a better way for modified nodal approach", *Proc. ECCTD*, pp. 864-872, 1991.

[30] G. Gielen, H. Walscharts, W. Sansen, "ISAAC: A Symbolic Simulator for Analog Integrated Circuits", *IEEE J. Solid-State Circuits*, vol. SC-24, no. 6, pp. 1587-1597, Dec. 1989.

[31] G. Gielen, *Design Automation for Analogue Integrated Circuits*. Ph. D. Dissertation, Katholieke Universiteit Leuven, 1990.

[32] G. Gielen, W. Sansen, *Symbolic Analysis for Automated Design of Analog Integrated Circuits*. Kluwer Academic Publishers, 1991.

[33] K. Gopal, M. S. Nakhla, K. Singhal, J. Vlach, "Distortion analysis of transistor networks", *IEEE. Trans. Circuits and Systems*, vol. CAS-25, no. 2, pp. 99-106, Feb. 1978.

[34] P. E. Gray, J. K. Matchett, "Comments on 'Pole and Zero Estimation in Linear Circuits'", *IEEE Trans. Circuits and Systems*, vol. CAS-38, no. 11, pp. 1404-1405, Nov. 1991.

[35] R. Gregorian, G. C. Temes, *Analog MOS Integrated Circuits for Signal Processing*. Wiley, 1986.

[36] S. B. Haley, P. J. Hurst, "Pole and Zero Estimation in Linear Circuits", *IEEE Trans. Circuits and Systems*, vol. CAS-36, no. 6, pp. 838-845, June 1989.

[37] S. B. Haley, P. J. Hurst, "Errata in 'Pole and Zero Estimation in Linear Circuits'", *IEEE Trans. Circuits and Systems*, vol. CAS-38, no. 11, p. 1406, Nov. 1991.

[38] S. B. Haley, P. J. Hurst, "Authors' Reply to 'Comments on 'Pole and Zero Estimation in Linear Circuits''", *IEEE Trans. Circuits and Systems–I: Fundamental Theory and Appl.*, vol. CAS-39, no. 5, p. 419, May 1992.

[39] C. W. Ho, A. E. Ruehli, P. A. Brennan, "The Modified Nodal Approach to Network Analysis", *IEEE Trans. Circuits and Systems*, vol. CAS-22, no. 6, pp. 504-509, June 1975.

[40] E. Hökenek, G. S. Moschytz, "Analysis of general switched-capacitor networks using indefinite admittance matrix", *Proc. IEE*, vol. 127, no.1, pp. 21-33, Feb. 1980.

Bibliography

[41] K. Kundert, A. Sangiovanni-Vincentelli, "Simulation of Nonlinear Circuits in the Frequency Domain", *IEEE Trans. Computer-Aided Design*, vol. CAD-5, no. 4, pp. 521-535, Oct. 1986.

[42] Y. L. Kuo, "Frequency-domain analysis of weakly nonlinear networks: "Canned" Volterra analysis, part 1", *Circuits and Systems*, pp. 2-8, Aug. 1977.

[43] Y. L. Kuo, "Frequency-domain analysis of weakly nonlinear networks: "Canned" Volterra analysis, part 2", *Circuits and Systems*, pp. 2-6, Oct. 1977.

[44] C. F. Kurth, G. S. Moschytz, "Nodal analysis of Switched-Capacitor Networks", *IEEE Trans. Circuits and Systems*, vol. CAS-26, no. 2, pp. 93-105, Feb. 1979.

[45] C. F. Kurth, G. S. Moschytz, "Two-Port Analysis of Switched-Capacitor Networks Using Four-Port Equivalent Circuits in the z-Domain", *IEEE Trans. Circuits and Systems*, vol. CAS-26, no. 3, pp. 166-180, Mar. 1979.

[46] B. Li, D. Gu, "SSCNAP: A Program for Symbolic Analysis of Switched Capacitor Circuits", *IEEE Trans. Computer-Aided Design*, vol. 11, no. 3, pp. 334-340, Mar. 1992.

[47] A. Liberatore, S. Manetti, "SAPEC - A Personal Computer Program for the Symbolic Analysis of Electric Circuits", *Proc. ISCAS*, pp. 897-900, 1988.

[48] P. M. Lin, "A Survey of Applications of Symbolic Network Functions", *IEEE Trans. Circuit Theory*, vol. CT-20, no. 6, pp. 732-737, Nov. 1973.

[49] P. M. Lin, *Symbolic Network Analysis*. Elsevier, 1991.

[50] J. G. Linvill, "Lumped Models of Transistors and Diodes", *Proc. IRE*, pp. 1141-1152, June 1958.

[51] J. Litsios, W. Fichtner, "Improved Expansion of Simplified Symbolic Determinants Using a Minimum Weighted Matching Algorithm", *Symbolic Methods and Applications to Circuit Design, Proc. SMACD'94*, Seville, Spain, pp. 231-239, Oct. 1994.

[52] J. E. Meyer, "MOS Models and Circuit Simulation", *RCA Review*, vol. 32, pp. 42-63, Mar. 1971.

[53] S. K. Mitra, "Equivalent Circuits of Gyrators", *Electron. Lett.*, vol. 3, no. 7, pp. 333-334, July 1967.

[54] G. S. Moschytz, *Linear Integrated Networks: Fundamentals*. New York: Van Nostrand Reinhold, 1974.

[55] B. R. Myers, "Nullor model of the transistor", *Proc. Inst. Elect. Electronics Engrs.*, vol. 53, pp. 758-759, July 1965.

[56] L. W. Nagel, *SPICE 2, A Computer Program to Simulate Semiconductor Circuits*. Technical Report ERL-M520, Electronics Research Lab, University of California, Berkely, May 1975.

[57] S. Narayanan, "Application of Volterra Series to Intermodulation Distortion Analysis of Transistor Feedback Amplifiers", *IEEE Trans. Circuit Theory*, vol. CT-17, no. 4, pp. 518-527, Nov. 1970.

[58] E. H. Nordholt, *Design of High-Performance Negative-Feedback Amplifiers*. Amsterdam: Elsevier, 1983.

[59] B. Pellegrini, "Considerations on the feedback theory", *Alta Frequenza*, no. 11, pp. 825-829, 1972.

[60] B. Pellegrini, "Some general considerations on the feedback systems employing integrated circuit amplifiers", *Alta Frequenza*, no. 8, pp. 442-445, 1972.

[61] P. Sannuti, N. N. Puri, "Symbolic Network Analysis - An Algebraic Formulation", *IEEE Trans. Circuits and Systems*, vol. CAS-27, no. 8, pp. 679-687, Aug. 1980.

Bibliography

[62] J. J. E. Slotine, W. Li, *Applied nonlinear control.* Prentice-Hall International, 1991.

[63] M. R. Spiegel, *Theory and Problems of Laplace Transforms.* McGraw-Hill, 1965.

[64] J. A. Starzyk, A. Konczykowska, "Flowgraph Analysis of Large Electronic Networks", *IEEE Trans. Circuits and Systems*, vol. CAS-33, no. 3, pp. 302-315, Mar. 1986.

[65] B. D. H. Tellegen, "The Gyrator, a New Electric Network Element", *Philips Research Reports*, vol. 3, pp. 81-101, April 1948.

[66] Y. P. Tsividis, *Operation and Modeling of the MOS Transistor.* McGraw-Hill, 1987.

[67] J. Vlach, K. Singhal, *Computer Methods for Circuit Analysis and Design.* Van Nostrand Reinhold, 1983.

[68] P. Wambacq, F. V. Fernández, G. Gielen, W. Sansen, A. Rodríguez-Vázquez, "Symbolic Network Analysis of Large Analogue Integrated Circuits", *Symbolic Methods and Applications to Circuit Design, Proc. SMACD'94*, Seville, Spain, pp. 185-209, Oct. 1994.

[69] D. E. Ward, R. W. Dutton, "A Charge-Oriented Model for MOS Transistor Capacitances", *IEEE J. Solid-State Circuits*, vol. SC-13, no. 5, pp. 703-707, Oct. 1978.

[70] S. Wolfram, Mathematica, *A System for Doing Mathematics by Computer.* Addison Wesley, 1991.

[71] Q. Yu, C. Sechen, "Approximate Symbolic Analysis of Large Analog Integrated Circuits", *Symbolic Methods and Applications to Circuit Design, Proc. SMACD'94*, Seville, Spain, pp. 241-259, Oct. 1994.

Index

A
Admittance matrix, 34, 36
a-parameters, 56, 59
Approximations
 symbolic, 37, 41, 80, 95, 121, 124
 examples, 6, 43, 45, 147

B
Bipolar transistor, *see* Transistor
Block diagrams, 53, 127

C
Capacitance matrix, 110
CASCA, 119
 examples, 6, 43, 45, 116, 125, 145
 graphics, 127
 language interpreter, 120
 subcircuit, 126
 subprogram (netlist), 122
 see also Approximations
Cascode
 CE-CB cascade, 68, 69, 71, 72
 differential amplifier, 69, 71, 73
CB-stage, 69, 73
CC-stage, 69, 72
CE-stage, 9
Chain matrix, 56
Charge equations, 110

Circuit simulation, 2
CMRR, 6
Cofactor, 34, 35, 52, 136
Common-mode gain, 6
Complexity, 34, 78
Constitutive equations, 18, 23
Controlled sources, 19
 VCVS, 49, 53, 127
Cramer's rule, 36, 136
Current mode circuits, 3

D
Describing functions, 84, 90, 94
 variable notation, 143
Design automation, 2
Determinants, 2, 33, 135
Differential equation, 15
Differential gain, 6, 44, 46
Differential pair, 63, 65
Driving point impedance, 37, 53

E
Element stamp, 50
Emitter-degeneration, 77
Equivalent analog circuit, 112

F
Feedback amplifiers, 57
 current amplifier, 59, 65, 67
 current follower, 69, 73
 gain, 57
 multi-stage, 9, 65

single-loop, 57, 62
transadmittance, 58, 65, 66, 68, 72
transimpedance, 12, 57, 65, 66, 68, 71
voltage amplifier, 12, 58, 65, 66
voltage follower, 69, 72
Feedback models
basic, 85
general, 91
Filters
active notch, 125, 146
SC-biquad, 116, 152
Fourier transformation, 92
Four-port network, 111
Functions
exponential, 101
nonlinear, 92
parabolic, 97
piecewise linear, 92, 106
polynomial, 16, 39, 122
saturation, 105
tanh, 11

G
Graph theory, 2, 30, 31
Gyrator, 20
nullor model of, 20, 21
transformations, 20, 49, 50, 51

H
Harmonic balance, 84, 88
Harmonic distortion, 83, 96
examples, 9, 12, 85
Horner's scheme, 40

I
Impedance, 37, 57
Incidence matrix, 24
Inversion, 135

J
Jacobian matrix, 90

L
Laplace
development, 136
transform, 15, 18
Large-signal analysis, 86
Long-tailed pair
CC-CB cascade, 9, 68, 70, 72
Loop equations, 27
Loop gain, 57

M
Mesh equations, 26
Minor, 136
Modified nodal equations, 29, 49
MOS transistor, *see* Transistor

N
Newton-Raphson, 40, 90
Nodal analysis, 36
compacted, 49, 51, 52, 109
see also Block diagrams
Nodal equations, 28
Node, 17
Nonplanar network, 26, 27
Norator, 19, 51, 59
Nullator, 19, 51, 59
Nullor, 3, 19, 51, 59, 109, 125
amplifier, 59, 77
balanced, 64, 65
buffer, 54
equivalence rules, 60, 61
gain, 59
gyrator model of, 20, 21
implementations, 55, 57, 63, 64, 69, 70
synthesis, 9, 55, 62, 65, 68
transconductance model of, 51
transistor, 59, 60, 77

O
Open circuit, 60, 61
Operational amplifier
    finite gain, 54, 113
    ideal, 113, 125
    single-stage, 45
    two-stage, 43, 139, 147

P
Permutation, 135
Poles, 16
    see also Roots
Pole-splitting
    extended, 39
Port, 17, 122

R
Roots, 122
    approximations, 41
    displacements, 38, 45, 47
    numerical, 39

S
Semi-symbolic, 2, 32, 33, 124, 127
Short circuit, 60, 61
Singular elements, 2, 19
Small-signal models, 19, 78
    see also Transistor
Switched-capacitor networks, 109
Symbolic analysis, 1

T
Tableau matrix, 26
Taylor series, 12, 92, 101
Term cancellations, 7, 29, 50
THD, *see* Harmonic distortion
Time-discrete analysis, 109
Transadmittance, 49, 50, 109, 123, 145
    see also Feedback amplifiers
Transcapacitance, 50, 78, 123
Transconductance, 50, 77, 123, 145
    balanced, 63
    nonlinear, 83, 85, 92
    see also Functions
Transfer impedance, 37, 53
Transfer voltage ratio, 37, 53
Transistor
    amplifiers, 55, 75
    bipolar, 78, 101
    compound, 63, 65, 67
    ideal, 59, 60, 75, 77
    large-signal, 83
    models, 75, 78
    MOS, 78, 97
    nonideal, 61
    nonlinear, 9, 61, 85
    two-stage pair, 63–67
Transmission equations, 56
Transmission function, 15
Transresistance, 19, 124
Transsusceptance
    inductive, 50, 109, 123
Two-port network, 17, 53, 56, 65, 76, 123

V
Volterra series, 84, 90, 105

Z
Zeros, 16
    see also Roots